# Super Stimulated

**Nicklas Brendborg** (b. 1995) is a Danish scientist and author. His international bestseller *Jellyfish Age Backwards* has been translated to more than 30 languages and was shortlisted for the Royal Society Science Book Prize. He is the youngest ever author to have been shortlisted for this award. Brendborg holds bachelor's and master's degrees in molecular biomedicine and biotechnology from the University of Copenhagen and is now a PhD student in molecular biology. He has been recognised by, among others, the Novo Nordisk International Talent Program and the Novo Nordisk Scholarship Program. The second book in his series of popular science books about longevity, *Super Stimulated* (*Vanedyr*) was a bestseller in Denmark in 2023 and is also being translated across the globe.

# Super Stimulated

## How Our Biology Is Being Manipulated to Create Bad Habits – and What We Can Do About It

Nicklas Brendborg

Translated by Sherilyn Nicolette Hellberg and Nicklas Brendborg

hodder press

First published in Great Britain in 2025 by Hodder Press
An imprint of Hodder & Stoughton Limited
An Hachette UK company

The authorised representative in the EEA is Hachette Ireland, 8 Castlecourt Centre, Dublin 15, D15 XTP3, Ireland (email: info@hbgi.ie)

2


First published in Denmark by Grønningen 1



A CIP catalogue record for this title is available from the British Library

Hardback ISBN 9781399735148
Trade Paperback ISBN 9781399735155
ebook ISBN 9781399735162

Typeset in Bembo by Hewer Text UK Ltd, Edinburgh
Printed and bound in Great Britain by Clays Ltd, Elcograf S.p.A.

Hodder & Stoughton policy is to use papers that are natural, renewable and recyclable products and made from wood grown in sustainable forests. The logging and manufacturing processes are expected to conform to the environmental regulations of the country of origin.

Hodder & Stoughton Limited
Carmelite House
50 Victoria Embankment
London EC4Y 0DZ

www.hodderpress.co.uk

# Contents

# Prologue: A small bird with a large egg

A small bird is sitting in a cage, glancing at a large egg. It briefly tilts its head as it considers what to do. And then it tries to hop onto the egg. It's quite an awkward scene. The egg is almost as large as the bird itself, so each time the bird tries to get comfortable, it slides right back down. Only to immediately try again.

A few metres away, a couple of Dutch scientists are observing the scene and chuckling knowingly at each other: *We've fooled this one too.*

One of the scientists is Nikolaas Tinbergen, who will go on to win the Nobel Prize for this experiment and others like it. And the bird being studied is called an oystercatcher. You might have seen one at the beach before. It has a black back and head, a white belly and trademark red legs and beak. Usually, oystercatchers lay small brown eggs that weigh less than 50 grams. But the Dutch researchers have discovered that the birds actually prefer eggs that are much larger than that. When the researchers make big fake eggs out of plaster, they can spellbind the poor birds, who always pick the exaggerated option instead of their own natural eggs.

Oystercatchers make this choice because they're instinctively attracted to the largest of their own eggs. In nature, egg size is a good indicator of general health and can help predict the likelihood of the egg hatching into a robust chick that can survive and reproduce. However, there are natural limits to *how* large an egg a small bird like the oystercatcher can lay. That means the bird has never developed an upper limit to its instinct. Its brain simply lives by the rule *the bigger, the better*.

Because of this, large plaster eggs are examples of what Tinbergen later dubbed supernormal stimuli, or *superstimuli* for short. They are exaggerations of what an animal is instinctively attracted to – a stimulus that is a bigger, brighter, or stronger version of the naturally occurring option.[1]

Superstimuli haven't only been used to fool oystercatchers, but many other birds as well. Take, for instance, the tiny songbird known as the pied flycatcher. These 10-centimetre-tall birds lay small eggs with a slight greenish-blue colour. Akin to the oystercatcher, the intensity of colour indicates how healthy the eggs are, so flycatchers tend to prefer the most brightly coloured ones.

But with a little blue paint, scientists can make plaster eggs that are much more vibrantly coloured than any flycatcher egg would ever be naturally. And just like the oystercatcher, this means pied flycatchers will pick the artificial superstimuli over their own eggs.[2] [3]

In fact, it turns out you don't have to be specifically creative to fool some birds. In one experiment, researchers simply placed some white volleyballs among a flock of geese. And then the geese left their own eggs to go and roost on the bigger, brighter volleyballs instead.

Birds are not the brightest, huh?

# Part 1

## FOOD SUPERSTIMULI

# 1

# The weight of modernity

You and I are living in the midst of a historic public health crisis.

You might not think about it much in your everyday life. And that's understandable. We quickly get used to our surroundings. But imagine we could travel back in time 150 years and bring a few confused people with us to the twenty-first century. These time travellers would be gobsmacked by the modern world: the cars on the roads, the planes in the sky and the abundance of food in our supermarkets. But they would also quickly notice that people look different now. We're taller than we used to be but we're also *wider.* And not just a little bit.

You see, it's not that excess weight or even obesity was completely unheard of in the past. But it *was* exceedingly rare. We know that from old books, paintings, clothes and also from actual measurements. Some of the first proper datasets of heights and weights for various people are from Britain in the mid-1800s. They are mostly measurements made on soldiers but there are also surveys of prisoners being sent to Australia and studies on the health of various social classes. In any case, having both heights and weights is interesting from a scientific standpoint. Together they allow us to calculate what is called body mass index, or BMI, which is the most commonly used method to classify whether someone is overweight. BMI gets a lot of criticism, and we will come to that. But let us first compare how modern people stack up against people from the past.

In the middle of the 1800s, an average 35-year-old British man had a BMI of around 23.[1] Today, an average British man of the same age has a BMI close to 28.[2] This difference might not sound like *that* much. But if we convert it into actual weights, it becomes quite concerning. For a man who is between 170 and 190 centimetres tall, this BMI increase is similar to a weight gain of between 14 and 18 kilograms. That is the equivalent of carrying around a large beagle or a medium-sized microwave oven at all times.

BMI critics might chime in here with an objection: BMI only takes body weight into account – not what that weight consists of! If the weight is excess fat, that is obviously a concern. But couldn't some of it be muscle? Arnold Schwarzenegger famously had a BMI of 31 when he was in peak condition. A BMI above 25 is classified as overweight and one exceeding 30 is classified as obese. But obviously prime Arnold Schwarzenegger was anything but obese.

Your average gym-goer might not have a physique quite as extreme as Arnold's. But it is true that having a lot of muscle mass will inflate your BMI above what it otherwise would be. However, only 16 per cent of the UK population has a gym membership[3] and when scientists actually scan people to measure their body fat percentages, it turns out BMI does not overestimate how many people are overweight. It *under*estimates it.[4] That's because the opposite of the Arnold Schwarzenegger phenomenon is common: many people have low muscle mass.

Thus they can carry lots of fat without weighing that much. Someone can have a BMI under 25 and be classified as normal weight, not because they are lean but because their low muscle mass gives them a low baseline weight. This is what is called being 'skinny fat'.

So, yes, on an individual level, BMI can get things wrong. But on a population level, BMI is accurate at judging body fat and if anything it underestimates the problem. That makes the current state of our health scary. According to BMI, *two thirds* of British adults are now overweight. And the same is true for Canadians, Australians and New Zealanders.[5] [6] [7] Even those numbers might be understating the problem, though.

In the modern world, people tend to gain weight as they age, meaning young people are the leanest. If we look only at the middle aged, between 45 and 65, a full *three out of four* British adults are overweight.[8] Think about that for a second. Here's a health problem that only one in four people manage to escape. Yet, despite the severity of the issue, things are still getting worse. Every single year waistlines expand a little more.

If we want to gauge how bad problems could get, we can look at the wider world. Because, of course, the obesity epidemic is not a phenomenon confined to the Anglosphere. The classic cautionary tale is the United States, where weight problems are even worse. But, despite the stereotype as the most overweight country in the world, the US does not actually top the charts anymore. Middle Eastern countries such as Egypt, Qatar and Kuwait are now heavier than the US. And at the very top of the list are a number of small island nations in the Pacific Ocean. 'Pacific islands' might conjure up paradise-like images of white sandy beaches, broad palm trees and colourful coral reefs. But the truth is paradise has been overcast by the heavy cloud of public health issues among the islanders. Practically speaking, almost no one on the Pacific islands qualifies as normal weight anymore. On most of the islands above 80 per cent of the population weigh too much, and for the record holder, Nauru, that number is closer to 90 per cent.[9]

So these former isles of paradise are like warning signs, flashing at the rest of the world: *No matter how grave your health issues might already be, they can still get worse.*

And that is exactly what is happening, no matter where we look. It won't be long before the global population as a whole crosses the dividing line of having more people who are overweight than normal weight. And the list of countries struggling with obesity spans the globe: Turkey, Mexico, Saudi Arabia, Chile, the Bahamas, Iraq, Belize, Malta, Israel, Hungary and so on.[10]

It's a list including every ethnicity, religion and climate type. Even in the poorest part of the world, Sub-Saharan Africa, excess fat is quickly becoming the norm. Sure, some regions still battle malnourishment and hunger as has historically been the case. But in many regions of Africa, there are now more overweight than underweight people.[11] [12] [13] And the most affected countries, such as South Africa, have already passed much of the developed world and are heading towards American numbers.[14]

So, it feels fitting to ask: what in the world is going on? Did we completely lose interest in our health?

Of course not. Quite the contrary, in fact. To our time travelling friends from previously, we would seem *obsessed* with our body weight. We are feverishly discussing calories and supplements while jumping on one fad diet after the other. Studies show that about half of all adults try to lose weight *every single year.*[15] But it hardly helps. Even those who manage to lose weight often end up putting most of it back on again. After a successful diet, the average person regains half the weight within two years. And within five years, that number is closer to 80 per cent.[16]

It's hard not to be left with a feeling that something is seriously wrong.

★

We have the same genes as our slim forefathers just a few generations ago, so genetics cannot be the root cause of the obesity epidemic. Some countries have even managed to go from slim to majority overweight in a single generation.

So, if the culprit isn't nature, it must be *nurture.* Something about our surroundings or our lifestyles must have changed drastically in the past few decades.

*Obviously!* you might think. In the past, most people worked physically demanding jobs all day. Now most of us sit at our desks digitally shuffling papers around. That clearly means we burn fewer calories than previously, which could explain our growing waistbands.

To test this theory, we need to go on a little detour to northern Tanzania. Here, we meet the Hadza people, who are some of the most physically active people in the world. The Hadza are hunter-gatherers, meaning their lifestyle is reminiscent of what ours used to be before the adoption of farming. They don't have crops or livestock, but rather set out on the savannah each morning to procure their food. The men hunt wild animals and climb for honey, while the women collect roots, berries, fruits and nuts.

In other words, the Hadza are much more physically active than modern Westerners. The average Westerner takes around 5,000 steps a day,[17] while it is 19,000 for Hadza men and 13,000 for Hadza women. Some of these steps are taken while dragging the daily spoils back to the community, whether they are root vegetables, guinea fowls or a whole zebra. It's no surprise, then, that the Hadza are in pretty good shape. As we learned previously, the average British person has a BMI of 28, putting them firmly in the overweight category and Canadians, Australians and New Zealanders are the same. The average Hadza, on the other hand, has a BMI of around 21. And this is while living an active lifestyle, meaning the average Hadza probably carries more muscle mass than the average British person for a given weight.

So, the explanation would seem simple: the sedentary Westerners burn few calories and are therefore overweight, while the active Hadza burn a lot of calories and are therefore slim and healthy.

The problem is, however, that this picture is incorrect. Scientists have done measurements and the Hadza actually *don't* burn more calories in a day than Westerners do, despite being vastly more active.[18]

If you're surprised, I can't blame you. I was too, the first time I heard this fact. But it's true. The Hadza don't burn more calories than Westerners.

In fact, if you just look at the raw numbers, the average Westerner burns more calories in a day than the average Hadza. This is because larger bodies require more energy and Westerners are often larger than the African hunter-gatherers.

But if you take body size out of the equation, the results become similar. So a group of equally sized Westerners and Hadza burn the same number of calories per day, even if the Westerners are desk-bound office workers while the Hadza are active hunter-gatherers constantly on the move.

The studies uncovering this surprising result were performed by the American evolutionary biologist Herman Pontzer and his colleagues. They used an advanced method called *doubly labelled water* to measure energy expenditure in humans and animals. We don't have to dive into the nitty gritty details, but rest assured that it is the most accurate means of measuring energy expenditure we have. Much more accurate than fitness trackers or other equipment that claims to do the same.

Herman Pontzer and co. have used the doubly labelled water method to measure energy expenditure in population groups around the world, from Siberian reindeer hunters to Latin American farmers. And in each case, it turns out that energy expenditure is roughly the same, at least on a population level.[19] There can be differences between individuals as some people burn more calories than others, even if they are the same size. But at the level of the *general population*, results are similar. Even for extremely active populations and populations that spend most of their day sitting still.

One of the best examples of Pontzer's research is a study that compared the metabolisms of two groups of children in Ecuador. Some were living in Amazonian tribes while the rest were modernised city-dwellers. In the cities of Ecuador, about a third of children are overweight, and you might be tempted to think this is because they don't move enough and therefore don't burn many calories. There's no doubt, after all, that city children are less active than children living in the rainforest.

But when the American scientists checked, it turned out the city-dwelling children burned just as many calories on a daily basis as the Amazonian children.[20] And the same amount as British and American children for that matter. In fact, Pontzer and his colleagues have even described this phenomenon among animals: chimpanzees, kangaroos and pandas in zoos burn just as many calories per day as their counterparts do in the wild, even though zoo animals are much less active.[21] [22] [23]

*

If you're feeling a bit confused right now, I understand. How can sedentary Westerners and active Hadza burn the same number of calories in a day? Being physically active takes energy, we know that much. So why isn't it visible in the total energy expenditure?

The reason seems to be one that we'll return to repeatedly in this book. Our bodies are *dynamic*. They are adaptation machines, constantly responding to their environments. In this case, in a pretty annoying way.

You see, your body has evolved in an environment of constant scarcity. It is attuned for conserving energy. This means that if you increase your daily energy expenditure, for example by exercising, your body tends to compensate by decreasing the amount of energy it expends elsewhere.[24] We can see an example of this in elite endurance athletes, who typically have lower levels of sex hormones like testosterone and oestrogen compared to people who don't exercise.[25] This is because elite endurance athletes can use so much energy on physical activity that their bodies begin to deprioritise other areas, such as reproduction. For female athletes who are especially active, this can even mean that they stop having periods.

The same phenomenon can be seen among hunter-gatherer communities. Not the absence of menstruation – hunter-gatherers have no issues reproducing. But just like elite endurance athletes, they tend to have lower levels of sex hormones compared to sedentary Westerners.[26] This might not sound appealing but consider that hormone-dependent cancers like breast cancer and prostate cancer are among the most common forms of this disease in the developed world, where we likely have much higher hormone levels than we evolved for.

Anyways. Reproduction isn't typically the first place the body looks to save energy. Typically, it will decrease energy expenditure in less important areas like unnecessary movement first. After exercise, you might be less active later in the day – get up and move around less, sit or lie down instead of standing, fidget less or refrain from maniacally bouncing your leg up and down in that way many of us are prone to do.[27]

Taken together, this means that the calories you burn by being physically active don't necessarily correspond 1:1 to a higher total daily energy expenditure. Maybe you burned 400 calories going for a run, but if your body subsequently decreases its energy expenditure in other areas, your total energy expenditure for that day might only end up being 250 calories higher than on a more idle day.

This sort of thing is precisely what we've seen in studies of people who exercise with the goal of losing weight. In one American study, researchers asked a few overweight young people to take up jogging with the aim of shedding a few pounds. The running programme was designed so that the participants would burn between 286 and 430 calories at each exercise session. But even though they kept to the training programme, their daily energy expenditure ended up being only about 220 calories more per day than usual. And so, while they lost weight, it was less than expected.[28]

Unfortunately, such tales are common for studies on exercise for weight loss. At the outset, study participants tend to burn as many calories per day as expected. But as more time passes, the body gets better at compensating for the increase in activity. And this is probably why the Hadza don't burn more calories than the British do. Their bodies have had their entire lives to learn how to compensate for their active lifestyles.

Don't get me wrong. I'm *not* saying that exercise is useless if you want to lose weight. As we saw in the American study, runners actually did end up increasing their daily energy expenditure a little. It's just that the data clearly supports the old adage: you cannot outrun a bad diet.

Of course, if you're stubborn enough, it is possible to override your body's attempt to compensate by sheer volume. If you start training like a Tour de France cyclist, for example, you'll burn so many calories that the body simply won't be able to compensate fully by decreasing energy expenditure elsewhere. Moreover, research on weight loss suggests that while exercise is often disappointing when it comes to shedding weight, it can be a deciding factor in not gaining weight back. And generally speaking, exercise is still among the healthiest things you can do. If you've read my book *Jellyfish Age Backwards*, you know, for example, that physical activity is one of the best ways to extend your life expectancy. It slows down the physical decline that comes with age and lessens the risk of most age-related diseases.

But with all those caveats out of the way, the original question remains. We visited the Hadza to learn whether Westerners have become overweight by being sedentary. And here, the answer is no. Because we burn just as many calories as populations that are significantly more active. Something else must underlie the obesity epidemic.

If you want more proof, we actually have good data on the physical activity levels of past decades too. There are multiple studies from the 1980s and 1990s tracking the daily activity levels of regular people. And these studies show that people back then were *less* physically active than we are today.[29] Yet, they were still much leaner.

So, back to the drawing board we go.

# 2

# Hijacking hunger

We have discarded 'genetics' as an explanation for the obesity epidemic. Because, as we established, our genetics are similar to those of our slim forefathers, yet we keep growing heavier. However, that doesn't mean genes cannot offer us *any* clues when it comes to obesity. After all, even though we're all living in the same obesity-promoting environment – whatever the cause – we are not all equally affected by it. In the past, we would all have been effortlessly lean but when moving into modernity, some of us grow obese, some grow overweight and some remain just as slim as our ancestors used to be. And *this* difference has a genetic component. For instance, take adopted children. Even though they're raised with their adopted families, their body weights as adults are actually more closely correlated to those of their biological family members. That is, their genes seem to predispose them to be sensitive to our obesity-promoting environment or not, regardless of how they grow up.[1] The same phenomenon is seen among identical twins. They stem from a single fertilised egg that split during early development. So identical twins are genetically identical. You could call them genetic clones of each other. In a few rare cases, identical twins are adopted by separate families, meaning the same genetics are exposed to two different environments.

But despite growing up in different families, identical twins tend to end up being surprisingly alike. And that is true of their body weights as well.[2]

To understand why, scientists have conducted what are called genome-wide association studies. These are huge studies where researchers examine the genes of hundreds of thousands of people and correlate them to various traits. For instance, if there are some genetic variants that are more common among tall people compared to short people, it might be because these variants increase your height. Similarly, if some genetic variants are often found in people with obesity, but not slim people, it might be because they increase the risk of becoming obese.

The big genome-wide association studies have already discovered several genetic variants that increase the risk of becoming obese in the modern world.[3] But they are not found in the kinds of genes that we might have naïvely expected. With no prior knowledge we might have guessed that the most important genes in promoting weight gain would have something to do with fat tissue or perhaps with metabolism or energy storage. But instead, the genetic variants that actually increase the risk of obesity have turned out largely to have something to do with *the brain.* The variants tend to influence traits such as reward processing, appetite, and impulse control.

The reason is that humans are not so different from our old friends the oystercatcher and the pied flycatcher from the beginning of the book. Yes, our brains are bigger and we're more intelligent. But we still fall for the same trick: *superstimuli.* The birds can be tricked by egg superstimuli – exaggerated versions of their own eggs that are unnaturally large or brightly coloured. We humans are tricked by *food* superstimuli. That is, artificial food that has been carefully designed to push all the right buttons in the reward system – the part of the brain that reinforces behaviours by making stuff feel good. How susceptible we are depends on our individual genetic makeup. A piece of chocolate might be rewarding to both of us. But if my brain reacts more strongly than yours, I will have greater risk of eating too much and becoming overweight. Just like a stronger reward from alcohol could predispose one to drink more and end up addicted.

**How much do we eat?**

Scientists have used various ways to estimate how much the average person eats today. For instance, some have looked at the total food production and subtracted estimated losses and waste.[4] [5] Using this method, it is evident that the average person today eats way more food than previously. In fact, the average calorie intake is so high that it is surprising the obesity epidemic isn't even worse. Our extra calorie intake alone is more than enough to explain our growing waistbands.

*

Imagine you're sitting on the couch on a random Tuesday night. Suddenly, an urge hits you. Isn't there a bag of crisps in the pantry? There's also some candy in there. But no, not tonight . . . Or?

You might imagine this battle is a simple fight that you're having with yourself. On one side is your reason and willpower and on the other side is that urge to eat 600 empty calories in a couple of minutes. The battle might be hard but at least it is fair. You versus yourself.

In reality, though, this conflict is anything but personal. In fact, thousands of other people are involved and, unfortunately for you, they're all playing on the opposite team.

This is because the company that makes the crisps and candy has one goal: that you *do* open that bag in the pantry. After all, food manufacturers work according to the same economic laws as any other company. They exist to make money for their shareholders. So that means your reason, your will power and your sense of fullness are some of the manufacturer's greatest enemies. If you don't open that bag in the pantry, more time will pass before you buy another one. The same is true if you eat only a little at a time because you get full or if your will power intervenes to stop you halfway through the bag. Thus, the crisp and candy manufacturer needs to fend off your self-control and ability to feel full if they want to make as much money as possible.

Now, you might think I'm being a little dramatic here. But you'd be shocked by the lengths that food manufacturers are willing to go to reach their sales targets. The largest food manufacturers have research parks around the world, where they employ hordes of well-credentialled researchers and set aside *billions* to spend on the careful optimisation of chips, cookies and microwave meals, as if they were life-saving medicines or tactical weapons.

Some of the questions these food scientists try to answer include: what's the perfect balance of fat and sugar? Can you change the structure of salt crystals to be maximally pleasant?

Will a cookie be more appealing if it is 2 per cent firmer or softer? Is E620 or E621 more stimulating for your appetite? Should those crisps be a slightly darker shade of yellow? And so on and so on.

These companies often produce thousands of varieties of the same product and use advanced scientific methods to examine how both mice and humans react to each. Some food manufacturers have even been caught using *brain scanners* to study how their products affect the reward system.[6]

It's hard not to chuckle a little at this image. Imagine a laboratory full of scientists in lab coats with serious faces and expensive high-tech equipment, toiling away to create the optimal frozen pizza. Okay, maybe tragicomic is a better description. Because despite how ridiculous it might sound, the truth is that the scientific optimisation of food works exactly as its manufacturers intend it to. I bet you've experienced it on your own body. It's that feeling when you grab a handful of crisps and you're suddenly powerless to stop. That isn't random. It's by design.

What the companies are doing is designing superstimuli. The whole thing involves carefully assembling various ingredients into a product that will make you hooked. For instance, an apple is the fruit of an apple tree. A steak is the muscle tissue of a cow. An Oreo, on the other hand, is an amalgamation of ingredients from multiple continents that have been carefully combined to create maximal reward in the reward system of your brain. An Oreo contains sugar that has been refined from sugarcane, white flour that has been milled from wheat, palm oil that has been extracted from palm trees, cocoa powder from cocoa beans, lecithin from soybeans to emulsify the mixture, various syrups for instance made from corn, lots of salt and a bunch of additives to add colour, lengthen shelf-life and enhance the flavour.

Every single ingredient of an Oreo has been processed in some way or another, and the Oreo only exists because of a long industrial process. For this reason, an Oreo is also what is called an 'ultra-processed food'. You'll typically be able to identify these products based on their long shelf-lives, colourful packaging, brand names and extensive lists of ingredients that are incomprehensible to the average person. Examples include candy, crisps, chocolate, ice cream, soda, energy drinks, fast food, ready meals, sausages and breakfast cereals.

Maybe you have already heard the term ultra-processed food in discussions on the cause of the obesity epidemic. This type of food has been increasingly under scrutiny, and for good reason. But it's important to point out that 'ultra-processed food' doesn't fully capture the problem we are facing. It's not that a food automatically becomes unhealthy or fattening because it has been in contact with a machine or processed in another way. We could easily imagine processing a food without making it into a food superstimuli. In fact, we could process foods to make them *healthier*, for instance by removing harmful fats, adding extra fibre or increasing the protein content. We already know of such ultra-processed foods that are health-promoting such as whey protein. This is the type of protein most often used in protein shakes. Whey protein is a component of milk and is separated out using industrial processes. Afterwards, artificial sweeteners and flavourings are added so whey protein is ultra-processed using any definition of the concept. But in studies the addition of whey protein to the diet tends to have *positive* effects on a range of health markers such as muscle and bone mass, even in elderly people. So the problem is not processing per se. It is making food superstimuli – processing foods with the specific aim of making them maximally rewarding to the brain to make you eat as much as possible. Eating too much causes weight gain and it is the excess weight – not the food itself – that is doing most of the harm.

★

If you want to see how good food companies are at their jobs, the American obesity researcher Kevin Hall and his colleagues have provided an excellent example.

In a careful study, the scientists recruited normal Americans to spend four weeks in a research centre where all their food was served by the researchers, who could then keep track of each person's intake. The participants were split into two groups. One group was served ultra-processed foods including frozen meals, breakfast cereals and packaged snacks. The other group was served unprocessed foods such as fresh meat, fruit, vegetables and rice.

The scientists tried to prepare the meals of the two groups as similarly as possible. There was always approximately the same number of calories on the two groups' plates and the researchers also ensured that each meal contained approximately the same quantities of protein, carbohydrates, fat, salt, sugar, fibre and so on. The only difference in the two sets of meals was that one group received food designed by food scientists while the other group received unprocessed food.

Over the next four weeks, the American scientists tracked the eating habits of their study subjects, regularly checking their weights and taking blood tests. And the results were revealing about why we're fighting an obesity epidemic. All the participants were told that they should just eat until they felt full and they wore loose-fitting clothes to make it harder for them to notice if they gained weight. However, the group eating food designed by food scientists consumed *500 calories more per day* compared to the group that was served unprocessed food.[7] In a later Japanese study with a similar design, the difference was even starker. Here, research participants eating ultra-processed foods ate *800 calories more per day*.[8] That means people eating foods designed by food scientists eat the equivalent of an entire additional meal before they feel as full as those eating unprocessed foods. Unsurprisingly, that meant the people eating ultra-processed food gained weight in both experiments.

In the American study, it was also noticeable that the group eating unprocessed food *lost* weight during the experiment. This makes sense given they were regular Americans who must be assumed to normally have high intakes of ultra-processed food in their daily diets. We can think of them as starting out somewhere in between the two extremes. When the participants switched to solely eating unprocessed food, their appetites began to normalise and this was not only seen at the weigh-ins but in blood work too. During the unprocessed diet, blood levels of the hunger hormone ghrelin fell, while the amount of the appetite-reducing hormone PYY increased. That's a testament to how ultra-processed food affects the body. Food companies have managed to engineer their products to manipulate the inner workings of our physiology.

Let's find out how they do this.

**Rats like cookies too**

In the early days of the obesity epidemic, researchers wanted to use laboratory rats and mice to study what was happening. But they were faced with a problem: how do you make rodents gain weight? Normally, these animals eat an appropriate amount and automatically stay lean. To cause obesity, scientists had to experiment. First, they tried force-feeding. It worked but took a long time and was obviously cruel. Next, they experimented with giving the animals food pellets that were extra high in fat. This worked better than force-feeding but still took a long time. Finally, however, some scientists found a simple solution. They began feeding their animals a bunch of our favourite supermarket snacks such as cookies, crisps, candy, sausages and crackers. In scientific circles, this kind of diet was dubbed 'the Cafeteria Diet' and it's proven to be so effective that most laboratory animals just can't stop eating and become obese in just a few weeks.[9]

# 3

# Tales of the white gold

In March 1493, two weather-beaten ships glided into the harbour of Palos de la Frontera in Spain's southwestern corner. On board was Christopher Columbus and a ragtag crew of Spaniards returning with spoils from their first expedition to the Americas. Captive Indigenous people sat silently on deck, foreshadowing the bloody history ahead. And in the hold were exotic plants and animals that the Spanish had carefully collected during their explorations. As the ship docked, the air was buzzing with excitement. The proud sailors jumped ashore and started displaying the strangest of creatures as crowds gathered around them. The sailors had pineapples and colourful parrots. They even had turkeys, which probably made some of the onlookers secretly fear that the devil himself was visiting them in some kind of bird disguise.

The voyages of Columbus marked the beginning of the European colonisation of the Americas with all the human fates that it has entailed: exploration, war, migration, slavery and nation-building. What is much less discussed, though, is the impact of all those other passengers – the plants and animals that began crossing the Atlantic as well. The Columbian Exchange, as this interchange is dubbed, spread European animals across the American continent. On the Plains of North

America, previously sedentary tribes of Native Americans adopted the horse and became the iconic horseback buffalo hunters. Meanwhile, sheep, pigs, cows, rabbits and many more spread to the furthest corners of the continent. Pigs especially tended to escape and adapt well to living on their own. Today this means that millions of wild pigs are roaming across the Americas.

In the Old World, the biggest impact of the Columbian Exchange was the flood of new American crops. Within a few centuries, national cuisines across Europe, Asia and Africa were upended. Italians learned about tomatoes for the first time. Indians began incorporating chillies in their curries. And on the British Isles an infatuation with the humble potato blossomed. Other good examples of American crops include corn, bell peppers, avocados, cacao, peanuts, squash, vanilla as well as the aforementioned pineapple. All were completely unknown to Europeans, Asians and Africans just a few hundred years ago.

A less-known plant that also crossed the Atlantic was a small bushy tree called the coca tree. At home in Peru and Bolivia, this tree had an important cultural role due to its leaves. If you chew coca leaves, you get a mild stimulant effect, which is traditionally used to prevent fatigue, for ceremonial purposes and to stave off everything from altitude sickness to hunger and thirst.

The coca tree didn't really catch on in Europe initially but that was the norm. Tomatoes were considered poisonous at first and it took royal endorsement and heavy pressure to make peasants start growing potatoes. Over time, however, the plants of America had a tendency to win over hearts and minds. For coca, the big change came in 1855, when German chemists started experimenting with the leaves and were able to isolate the active compound. That is, they discovered which chemical is responsible for the energy-boosting effect and managed to extract it.

The chemists named the new substance after the coca tree – *cocaine* – and from here you might suspect the story diverges slightly from that of potatoes and tomatoes.

You see, initially the German chemists had planned to search for medical uses for their new substance. Cocaine did in fact end up in different medicines in its early years, but it didn't take long before people also found other uses. As it turns out, pure cocaine is a vastly stronger stimulant than cocaine bound up in coca leaves. There are two reasons for that. First, coca leaves are only 1 per cent cocaine, so it would take an elephant-like level of leaf consumption to approach the intake that people easily get from pure cocaine. And second, pure cocaine has a stronger effect than coca leaves because it takes less time to reach the brain. When you chew coca leaves, you're not only consuming cocaine, but also lots of fibre and water, which delays uptake into the body. Ultimately, this delay means it takes longer for the cocaine to reach the brain. And this is important, because we know that substances are generally more rewarding – and addictive – the quicker they're able to reach the brain.[1] Cigarettes, for example, are more addictive than nicotine gum because nicotine reaches the brain more quickly when you smoke it. In cigarettes, the nicotine is absorbed through the lungs, which is quicker than absorbing it through the mucous membranes of the mouth, as you do with nicotine gum. This is also why cocaine can be made even more addictive by transforming it into 'crack cocaine' that is smoked instead of taken through the nose.

Okay, maybe you are wondering what cocaine has to do with any of the things we've been discussing so far. I don't want to make you *too* paranoid about the food in your fridge, but there's an important parallel to food superstimuli here. Food manufacturers use the exact same trick to make food superstimuli as the German chemists used back then to develop cocaine: identify the pleasurable compound, isolate it and add tons of it in concentrated form. The best example of this process in action comes from a white powder that crossed the Atlantic in the opposite direction.

*

Before reaching the New World, Spanish and – especially – Portuguese explorers had spent a century sailing further and further away from home. The sailors ventured down the west coast of Africa and eventually also straight out into the Atlantic Ocean. During these voyages, the Spanish and Portuguese found several islands, many of which were uninhabited at the time: Madeira, the Canaries, the Azores, the Cape Verde islands and São Tomé & Príncipe. The respective monarchs were quick to claim the newfound islands for the crown and sent out settlers on a mission to make the islands profitable. A few different strategies were tried. But it didn't take long before some of the newcomers realised that the Atlantic islands were well suited for cultivating sugarcane – the crop that is processed to produce refined sugar.

Europeans had known about sugarcane since crusaders returned from the Middle East telling stories about delicious 'sweet salt'. But sugarcane needs a warm climate to grow so European production was naturally limited. Only a few small pockets at the very south of the continent were suitable to grow the crop and refined sugar was expensive enough that it gained the nickname 'white gold'.

Using their newfound islands to grow sugarcane, the Spanish and Portuguese thus managed to turn a nice profit. But even though they increased European sugar production, none of the islands on which it was grown were particularly big. For now, sugar remained a luxury product only affordable to the elite.

All this began to change when Columbus and his crew docked at that harbour in Palos de la Frontera with tales of the New World. Here, Europeans found boundless land that was perfect for growing sugarcane, and settlers spared no time in heading there. They had heard the stories of men amassing fortunes by growing sugar on the islands of the Atlantic and wanted in on the spoils. Some immigrants even came straight from these very islands, hoping to strike it even bigger in the Americas. Thus, just nine years after Columbus came ashore, sugarcane was harvested on the American continent for the first time. Before long, sugar had become one of the continent's key exports. And the following centuries saw a fierce battle commence to increase production and control the market. Sugarcane farming is both dangerous and labour intensive, meaning the colonists had a hard time hiring. To meet the demand and fill their pockets, they turned to slavery. That is obviously appalling wherever it occurs, but slavery was especially grim on sugar plantations. Enslaved people here had the worst fate among any brought to the Americas. Between 5 and 10 per cent of the workers died each year and the brutal conditions meant the workforce had to be constantly replaced.

Besides using slavery, the sugarcane growers also cut costs by inventing a host of new methods and technologies to improve their yield and for extracting as much sugar as possible from the cane. This optimisation process led to sugar prices falling drastically and as sugar became more available, more people started eating it. Eventually, prices dropped all the way to the floor, when German researchers bred what we now call sugar beets. These beets provided an easier way to obtain refined sugar compared to sugarcane as they don't require the heat of the tropics and could be grown in Europe close to the consumers. At this point, sugar had turned from 'white gold' into one of the cheapest sources of calories in existence. And so it remains to this day. Depending on where you live, a few staples such as rice or flour might be slightly cheaper. But almost all other foods today are more expensive per calorie than refined sugar.[2] Coupled with the attractive taste, the low price of sugar means that we now eat *a lot*. The average British person consumes around 70 grams of added sugar daily[3] while the average Canadian eats 85 grams.[4] That is more than a tenth of all the calories consumed, or equivalent to eating 18 teaspoons of pure sugar every single day.

★

Humans are not the only animals that love sugar. A few, such as dolphins and various cats, can't taste sweetness because they don't have functional taste receptors for it anymore. But besides that most animals have a sweet tooth. The reason is that sugar is an easy calorie source that doesn't require much digestion. For instance, there are 4 calories in both 1 gram of sugar and 1 gram of protein. But digesting and using protein is so energy intensive that it requires 15–30 per cent of the calorie content.[5] Sugar, on the other hand, doesn't take much digestion, meaning it is a quick fuel source. That's why athletes are fond of consuming simple sugars when they need energy.

Another reason humans in particular like sweet flavours is that we have a long evolutionary history as fruit-eating apes in the rainforest. In fruit, sweetness is an indicator of ripeness and tells you that the fruit isn't poisonous. Hence, for our ape ancestors millions of years ago, a sweet tooth helped in choosing the best and safest food options.

Today, fruits are still popular, as you can witness in any modern supermarket. Yet, modern fruits are far from what our ancestors ate. Take peaches, for example. Thousands of years ago, peaches weren't the juicy, apple-sized fruits of today. Wild peaches were the size of berries and had a dry and dull flavour sort of like lentils. However, in eastern China, farmers took an interest in peach trees and began to cultivate them. When planting the next generation, the farmers used stones from their best trees. This meant that the peaches were gradually selected for satisfying the taste preferences of the farmers – they became bigger, juicer and sweeter.[6] [7] [8] The same story could be told about pretty much all our other fruits. For instance, the predecessors of modern-day watermelons were small berry-like fruits with a bitter, green flesh.[9] And wild bananas don't remotely resemble the sweet fruits we know today. They're small and bitter with hard pits.

Don't get me wrong – the fact that our fruits are sweeter than 'natural' fruits doesn't make them unhealthy. On the contrary, fruits are healthy for humans: they're a good source of vitamins and minerals and they contain fibre which makes us feel full. But fruits are still the first example of something we will see again and again in the coming chapters: humans changing our foods to fit our tastes. And while this trend might have started out fine, it has definitely not remained so.

In the modern world, we can make foods sweeter using refined sugar. That has let us dial up the sweetness to unprecedented levels. The simplest example is dissolving some sugar in water and adding flavouring. You don't need more than that to create billion-dollar industries such as soda, iced tea, juices and energy drinks. The average British person drinks 117 litres of these drinks every year and it is the largest source of added sugar in the British diet. That's not only a problem because it keeps dentists around the country up at night. It's also a problem because sugary drinks are one of the most efficient ways to fatten up a human.

These beverages simply don't produce much satiety.[10] As we have discussed, our bodies are normally good at balancing our calorie intake. Your natural appetite control keeps track of how much you have eaten and adjusts your hunger accordingly. So, if you eat a bigger lunch than normal, you will tend to eat less at dinner. However, sugary drinks evade this process. In experiments where people are presented with a sugary drink, it doesn't actually decrease how much food they eat at a meal.[11] The drink simply adds extra calories. Why this happens is not completely clear but there are a few potential reasons. First, the high sugar content of a soft drink is a superstimulus that makes the reward system in the brain scream for more, even if other parts of the brain try to put on the brakes. Second, most soft drinks don't contain any of the substances that normally make us feel full such as protein and fibre. And third, liquids pass more quickly through the stomach. There are so-called stretch receptors in the stomach that register how full it is. Once room runs out, they send nerve signals to the brain to make you stop eating. However, while liquids fill up the stomach, they quickly empty into the intestines, meaning fullness dissipates quickly.

Whatever the reason, people tend to increase their calorie intake when drinking sugary drinks. That means these beverages make people gain weight. This is also evident the other way around. People who quit soft drinks tend to lose weight. In both cases we also see the expected downstream health effects such as on cholesterol levels and blood pressure. For instance, in a Swiss study, researchers asked a group of people who didn't typically drink soda to start to drink it. During the experiment, the researchers took regular blood tests, and after two to three weeks they could already see various health markers going in the wrong direction. Cholesterol levels rose, as did the blood pressure and inflammation markers of the subjects.[12]

★

It probably won't surprise you that sugar is also a cornerstone of other food superstimuli. Ice cream, candy and chocolate all get their appeal at least in part due to copious amounts of added sugar. Interestingly, the saga doesn't stop with sweet foods, though. Sugar isn't just the secret behind the appeal of snacks and desserts. You'll find it added to practically everything in the supermarket.

When I do talks, I usually bring along a little quiz where one of the questions is: 'Which of these products doesn't contain added sugar?'

- Cold cuts
- Wholewheat bread
- Canned kidney beans
- Cigarettes
- Pasta

No crowd ever gets it right. Most often they answer 'cigarettes' – but I can assure you, if there's one industry that knows how to exploit *every* single weakness of the human mind, it's the tobacco industry. The tobacco industry has long learned that adding sugar to cigarettes makes the smoke tastier. That makes people smoke more and fill up the coffers of tobacco companies.

No, the correct answer to my question is 'pasta'. All the other products typically have added sugar, even if they don't taste like it. That's because you don't just register the content of your meals by tasting them in your mouth. Most animals have special nerve cells in the intestines that monitor the food that passes by. Some of these nerve cells register sugary substances and notify the brain.[13] *Whatever you ate earlier was good, do it again if you get the chance.* We can see this play out in studies on rodents: if you give rats infusions of liquids directly into the stomach, they obviously can't taste them. But they still end up preferring the ones with added sugar.[14]

Now, the small amount of sugar in a piece of wholewheat bread or some canned kidney beans shouldn't frighten us. We're talking low amounts and foods that also come with plenty of fibre, protein, vitamins and minerals.

Instead, you should mostly take this example as an illustration of the appetite-stimulating power of sugar. It is so great that food companies almost can't help themselves adding just a little to their products. As a result, there are also large amounts of sugar in the non-sweet foods that we *should* cut down on. Fast-food chains, for example, are well aware of what the white gold does to our appetite, and they add sugar to practically everything on their menus. There's added sugar in pizza dough and tomato sauce, in burger buns, in dressings, sushi rice, Thai food, Indian curries and so on. In a tragicomic example, the Supreme Court of Ireland has ruled that the sandwich bread served by fast-food chain Subway has so much sugar that it cannot legally qualify as bread.[15]

**Undercover sugar**

Some food companies list 'sugar' as an ingredient but this is not always the best sales tactic. Instead, many try to get around it by using ingredients that sound like something else but are actually just sugar. These include 'dextrose', 'glucose', 'sucrose', 'fructose', 'agave nectar', 'evaporated cane juice' and various 'syrups'.

The sneakiest way to keep sugar off the ingredients list is to use different kinds of 'concentrate'. If, for example, you were to see 'fruit concentrate' on a list of ingredients, you probably wouldn't think too much of it. But fruit concentrate is essentially just sugar. It is made by first peeling the fruit, which removes the majority of the vitamins and fibre. Then, juice is made from what is left while removing the pulp that has the remaining fibre. Finally, the water is evaporated, leaving not much else but sugar. This trick is often used in packaged snacks that are targeted at health-conscious consumers, because 'fruit concentrate' sounds healthier than 'sugar'.

★

There are several diseases tied to sugar intake. A well-known one is that sugar promotes cavities in teeth. This happens because sugar feeds the bacterium *Streptococcus mutans* in the mouth which produces acid that wears down teeth. Besides that, the type of sugar called fructose is harmful to the liver in high doses such as those coming from sugar-sweetened beverages. But most of sugar's ill effects on health are actually more indirect. They stem from the simple fact that sugar is just too tasty. There's pretty much nothing else that can stimulate human appetite like adding sugar to a food and that makes us overeat. In turn, we gain weight and it is the excess body fat that is responsible for many of the health issues that can arise from sugar intake: increased risk of diabetes and cardiovascular diseases, chronic inflammation and general metabolic dysfunction. So it is not sugar that is the poison – it is the excess calories that it makes us consume. Perhaps there's a solution to that.

All the way back in 1879, Russian chemist Constantin Fahlberg spent one morning in the laboratory working on coal tar derivatives. At lunchtime, he went out to eat a sandwich he had brought from home but puzzlingly it didn't taste as expected. The sandwich was sweet. Fahlberg picked up a napkin to wipe his mouth but the napkin was even sweeter. He was confused but then it dawned on him: he had forgotten to wash his hands. The sweet taste had to come from one of the chemicals he was working with. As one does, Fahlberg hurried back to the laboratory and started tasting his reagents one by one. Fortunately, for his health, it didn't take long before he hit on the source – a molecule we now call saccharine, which was the first artificial sweetener.[16]

This story might sound a little crazy but it is pretty much the story of how *all* artificial sweeteners were discovered. For instance, the most common artificial sweeteners today are aspartame and acesulfame-K. Aspartame was originally discovered by American chemist James Schlatter who was doing an experiment and absentmindedly licked his fingers to turn a page in his laboratory book. And acesulfame-K was discovered by German chemist Karl Clauss who was in the laboratory, noticed he had a spot on his sweater and licked his finger to remove it.[17]

All these molecules have in common the fact that they aren't sugar but still taste sweet. That's because they activate taste receptors on your tongue that sense sweetness. Think of it this way: the taste receptors are like locks designed to be opened by specific keys in the form of sugars. And artificial sweeteners are molecules that just happen to have shapes that fit the lock even if they aren't what it's designed for. It's like someone trying millions of keys on your front door until finding one that, by sheer chance, fits and turns the lock.

In both cases, activation of the taste receptors triggers a nerve signal that travels from your tongue to your brain, where you sense it as sweetness. But here's the clever thing: artificial sweeteners either have few or no calories. Many are substances that we cannot metabolise, meaning they pass through our system unchanged without providing energy. Others do get metabolised but are used in tiny amounts, meaning they provide negligible calories. This is interesting, given how excess calories are responsible for so many of the ill effects of sugar. We can see the difference in the Swiss study from earlier. Remember? Researchers recruited people who normally didn't drink soda and had them begin doing so. Within two to three weeks the researchers could see all kinds of health markers going in the wrong direction: cholesterol levels rose and so did both blood pressure and inflammation markers. In a similar study the researchers also made some non-soda-drinkers start drinking diet sodas (which are sweetened with artificial sweeteners). This worked out quite differently. When non-soda drinkers begin drinking diet soda, nothing happens to these health markers, or to any others for that matter. The diet soda drinkers actually had health markers comparable to those who stuck to water throughout the study.[18] In other studies, researchers have also investigated what happens if you make regular soda drinkers switch to diet options. This improves their health markers and body weight. In fact the improvement is as great as when the researchers make soda drinkers switch to drinking water only.[19] [20]

Now, obviously these studies are not enough to conclude that diet soda is as healthy as water. But what we *can* conclude is that diet options are vastly healthier than the sugar-sweetened options. Some might object here – haven't artificial sweeteners had a lot of bad press? And that's true. Aspartame in particular is often under scrutiny but it is based on bad science, often isolated studies on cell cultures or rodents using huge amounts.[21]

Instead, let's follow what happens when you ingest aspartame, for instance by drinking a diet soda. The first time the aspartame interacts with your body is in your mouth. As we discussed, it activates taste receptors on your tongue, making you taste sweetness. Then it is swallowed, passes through your stomach and goes to the intestines where it is digested by enzymes, just like most other things you eat. Digestion turns aspartame into three different molecules: aspartic acid, phenylalanine and methanol.[22] These are the molecules that are actually taken up by your body, not aspartame itself. That means aspartame itself could only theoretically cause problems in your digestive tract. Any ill effect inside your body would have to come from aspartic acid, phenylalanine or methanol. So let's follow what happens to those three after entering the body, one by one.

First, aspartic acid is an amino acid that your body can produce on its own and is also abundant in practically all foods. The aspartame from a single diet soda will yield around 80 mg of aspartic acid. On any normal day, the average person eats around 6,500 mg, meaning a diet soda would account for a little more than 1 per cent. So there is simply no world where the aspartic acid from aspartame ever causes you any problems.

Next, phenylalanine is another amino acid. This time, you do not produce it yourself, so phenylalanine is what's called an essential amino acid. You need to get it from your diet to survive. Fortunately, this is trivial as phenylalanine is also abundant in practically everything you eat. A diet soda yields 100 mg of phenylalanine but on average you get 3,400 mg a day from other food, so the phenylalanine from a diet soda is around 3 per cent of that. Again, this just cannot cause you health problems. If you needed any more proof, a single glass of milk contains 13 times more aspartic acid than a diet soda and six times more phenylalanine.

That leaves us with the final breakdown product: methanol. This is the only compound that could theoretically harm your body. Methanol is an alcohol and, just like its cousin ethanol, it is toxic in large quantities. The danger lies in how methanol is broken down in the liver: it's converted to something called formaldehyde, which can damage proteins, DNA and other molecules. However, the amount of methanol you get from a diet soda is tiny – around 20 mg per can. In comparison, a single glass of orange juice has two to three times that amount. Even an apple has more. Generally, all fruits and vegetables contain a little methanol and that is completely okay. Our livers easily deal with it in low amounts. If you wanted to consume levels of aspartame that could actually harm you through the methanol content, you would need to drink at least 14 cans every single day. That number already includes a conservative safety margin so the actual limit is probably higher.

The other common artificial sweetener, acesulfame-K, is even easier to account for. It is not digested but taken up as is. Then it flows around the bloodstream for a while before being excreted in the urine. While it is in the bloodstream it doesn't seem to interact with anything. At least there are no effects on any blood markers.

Many other artificial sweeteners are the same, though not all. Some don't get taken up by the body at all, meaning they reach the colon where the majority of our gut bacteria live. This can cause changes in the abundance of various bacterial species, which might be detrimental, and which can also cause an upset stomach.[23] However, even in these cases, the evidence points to artificial sweeteners being a preferable option to the large amounts of refined sugar they often replace. Of course, water or the absence of sweetener is even better in most cases.

My experience is that this fact makes some people irrationally angry. Perhaps it's the 'artificial' part of the name that make some people automatically assume artificial sweeteners must be bad. But while it is not always correct to equal naturalness with healthiness, I can also assure you that there is absolutely nothing natural in downing literal liquid candy.

The opposition to artificial sweeteners might also be that inner Puritan whispering that pleasant things are always bad for you. Sometimes, though, there is such a thing as a win-win situation. And this is an important lesson to keep in mind for the rest of this book. The point is not that pleasure is always bad and will kill you if you succumb to it. No, the problem this book tries to highlight, is that strongly rewarding substances and experiences have a tendency to knock out our natural control mechanisms, causing overconsumption. If we can avoid the overconsumption, all is fine. Or alternatively, if we can identify the negative consequences of the overconsumption and eliminate them, that works, too.

**Confusion about calories**

Calories are a measurement of energy. The definition is that a calorie is the amount of energy it takes to raise the temperature of 1 gram of water by 1°C at atmospheric pressure. However, the way we are using the term in this book is technically wrong. So is the way we use it in daily life. The 'calories' on food labels and the ones we discuss here are in fact 'kilocalories'. That is, they are actually thousands of calories, just like kilometres are thousands of metres. Hence, an apple doesn't contain 80 calories, it contains 80,000 calories. That doesn't matter, though, because your energy needs are also not 2–3,000 calories, they are 2–3,000,000 calories.

Some people try to fix this misunderstanding by writing 'kcal' while others try to at least indicate that they are aware of the problem by capitalising the word as Calories. In this book, I have chosen to continue the confusion by just saying 'calories', in the hope that the world can forgive me.

# 4

# Rabbit starvation

During World War II, the United States Army distributed a survival guide to its air crew which included this warning: 'Under no conditions limit yourself to a meat diet of rabbit just because they happen to be plentiful in the region where you are forced down. A continued diet of rabbit will produce *rabbit starvation* – diarrhoea will begin in about a week and if the diet is continued, DEATH MAY RESULT.'[1]

'Died from rabbit starvation' doesn't sound very cool in a eulogy, so consider yourself warned.

All right, you don't have to be scared if someone offers you rabbit ragu. Rabbit meat is neither poisonous nor unhealthy. You just can't live *exclusively* on rabbit. This is because humans need a bit of fat in our diets and rabbits are simply too lean.

You see, like other animals, rabbits aren't particularly interested in being eaten. However, they don't have horns, sharp claws or towering body sizes to fend off predators. Instead, the survival strategy of rabbits is to be very difficult to capture – they have eyes on the side of their head to get a nearly 360-degree field of view. They have huge ears that register even small sounds and that can turn around to catch the exact direction. And finally, rabbits are quicker and more acrobatic than the vast majority of animals. They can change direction in an instant and jump several times their body height. All these acrobatic abilities require rabbits to be skinny, though. Each excess gram of fat could be the difference between barely getting away and ending up as someone's dinner. So rabbit meat has 1-2 grams of fat per 100 gram serving. This makes rabbits one of the leanest animals in the world, but other wild animals are also surprisingly slim. For instance, you could also be struck by rabbit starvation if you lived exclusively on reindeer during certain parts of the year.

The lean animals of the wild stand in strong contrast to the animals you and I eat in our daily lives. It might not be immediately apparent, but our livestock are quite hefty. Sometimes this is due to having unhealthy diets and miserable living conditions. But even when raised in fresh air and with room to move, our livestock carry quite a lot more body fat than their wild predecessors. Pigs, for example, are fatter than wild boars, even when raised outdoors.[2] And cows also carry more body fat than their close relatives, bison.[3]

Does any of this ring a bell? If not, here's a hint:

Peaches.

Just like our taste for sweetness has turned peaches from lentil-like to big, juicy and sweet, our innate preferences have shaped our livestock. Ever since we began domesticating various animal species, we've been selectively breeding the individuals that have characteristics we like. This has given us vastly more peaceful animals that are easier to control than their wild cousins. Just think about the difference between a docile pig and an aggressive wild boar. But during domestication, we have also changed the body composition of livestock according to our tastes – in this case according to our love of fat.

We have a soft spot for fat because fat is the most concentrated source of energy among the three macronutrients in our diet: fat, protein and carbohydrates. Both carbohydrates and protein contain 4 calories per gram whereas fat contains 9. That made fat valuable to our ancestors who practically all lived with a constant danger of starvation hanging over them. Imagine, for example, two Stone Age people. One of them is crazy about nuts, while the other one loves celery. Nuts are high in fat and therefore a concentrated source of calories, whereas celery is not. In fact, it would take a whole kilogram of celery to get the same amount of calories as you can get from a small handful of nuts. Thus, as a Stone Age person you would really shoot yourself in the foot by picking celery over nuts. Doing so would put you at an increased risk of starvation and any descendant of yours who inherited the trait would also not fare well. On the contrary, searching out the most concentrated calorie sources was a winning strategy for our ancestors. Besides keeping starvation at bay, it also allowed them to use all the extra calories for useful stuff like growing ginormous brains. So here we are liking cheese (as do many other animals by the way).

The actual physiological mechanisms behind our liking of fat are a little more opaque than those involved in our sweet tooth. 'Fat' is not one of the five official flavours – sweet, sour, salty, bitter and umami. But recent science suggests we might have taste receptors for fat too.[4] In any case, food scientists agree we also register the presence of fat in our mouths through what's called 'mouth feel'. Think about the difference between porridge made with milk and porridge made with water. The one with milk has a richer, creamier texture that feels more satisfying in the mouth but the reason why is not perfectly clear yet.

Our attraction to fat is not only a matter of what happens in the mouth, though. Just as we've already discussed for sweetness, there are neurons in our guts that sense the presence of fats in the food we eat and tell the brain about it.[5] That's why it's difficult to trick the body with low-fat diet foods. Manufacturers often succeed in mimicking the 'mouth feel' of fat with ingredients that have a similar texture. But the neurons in the gut aren't so easily fooled. They register that the diet version has less fat and fewer calories. When this information is relayed back to the brain, it means that we nevertheless end up preferring the fatty version, even if the two dishes are otherwise identical in taste.[6]

★

Fat is used in most forms of food superstimuli. But one category that many people especially think about when it comes to fat is fast food. Just consider a fast-food burger: fatty meat, fatty cheese, fatty bacon and fatty sauces.

I won't be defending fast food chains in this book – they're undoubtedly good examples of using food scientists to manipulate customers into overeating. But the fixation on fast food in some circles means that other parts of the food industry are allowed to go relatively unnoticed while using the exact same tricks. For instance, in one large study, an international group of researchers compared the caloric content of fast food and food from normal restaurants. You know, the ones where you sit at a table and are served by a waiter. If you're like most people, you probably assume that the food at restaurants is healthier than the food at fast-food chains. At least that's how I used to think. At a restaurant, there's usually some vegetables on the plate, and the food also tends to *look* less fatty. But ironically, the big international study showed that most restaurant meals have *more* calories than a typical meal at a fast-food chain.[7]

This is because restaurants, just like fast-food chains, are in bitter competition with each other. And it's not easy winning this competition if you refuse to use all the weapons at your disposal. If a restaurant wants us to come back, it must compete on all aspects of taste, including 'cheap' ones, like using large quantities of fat. If you know a restaurant chef, just ask them how much butter they use to cook a steak. That's why it tastes better than at home. Of course, restaurants also add sugar to make their food tasty. But it turns out, using extra fat is especially clever. Studies show we're typically pretty good at evaluating the sugar content of meals when blind tested. But our evaluations of the fat content are mostly way off.[8] Researchers can double the amount of fat in a dish without their study participants noticing any difference – consciously at least. As we've discussed, the gut will keep the score and makes us attracted to the fatty dish.

Besides, 'add fat' and 'add sugar' are complementary tricks. When sugar is added to a food without changing the level of fat, people start judging that food as *less* fatty than before.[9] That is, you can make a food feel less 'fatty' by just adding sugar. This trick also helps because the combination of fat and sugar is intensely pleasurable for both animals and humans. It is the magical formula behind pretty much all our favourite snacks. For instance, in addition to sugar, ice cream contains fat from cream. Chocolate contains fat from chocolate beans. Cake contains fat from butter or oil, and so on. We don't know for certain what makes the combination of fat and sugar so irresistible. In fact, it's practically never seen in nature. You'll find natural foods that are high in sugar, like honey. And you can also find natural foods that are high in fat, such as nuts. But you just don't see both at the same time.

The reason this combination works so well might simply be that humans love combining different pleasurable stimuli. We eat candy while watching a movie, drink alcohol while dancing to music and centre our socialising around eating good food. However, there is one other possible explanation for why we like combining fat and sugar. And now, I'm sorry if I'm about to ruin your favourite snacks for you. But it was not entirely true when I said *no* natural foods combine high levels of fat and sugar. There is one that does exactly that . . .

Breastmilk.

In breastmilk, about 55 per cent of the calories come from fat, 40 per cent from carbohydrates (in the form of lactose, which is also called milk sugar) and 5 per cent from protein.[10] This matches all kinds of supermarket snacks suspiciously well.

Let's take a few examples and list them by the percentage of calories from fat:carbohydrates:protein:

- Breastmilk: 55:40:5
- Chocolate: 50:45:5

- Ice cream: 50:45:5
- Danish pastries: 50:45:5
- Doughnuts: 50:45:5
- Cheese puffs: 55:40:5
- Crisps: 55:40:5

In some of these instances the carbohydrate source is mostly sugar (just as in breastmilk) but sometimes it is starch, for instance in crisps, or a mix like in various cakes. In any case, this specific caloric ratio is not found in nature outside breastmilk. Kidney beans, for example, are 5:70:25 and steak is 50:0:50.

Now, I'm not saying food scientists have specifically tried to hit the caloric ratio of breastmilk. They have simply been optimising the palatability of their foods in relative isolation and ended up with the same result. When perfecting food superstimuli, scientists call this searching for 'bliss points' – combinations that are maximally pleasurable for the laboratory animal or human research participant. Diverging from the bliss point ratio in any direction (more fat/less carbohydrate or less fat/more carbohydrate) will make the food less appealing. We can only speculate about why this ratio works so well but here's a possibility: think about the time in your life where you'd gain the most by having a voracious appetite. It's not as an adult where excess calories tend to get stored as fat. But as a baby, when a large appetite is necessary. Babies burn 50 per cent more calories than adults when adjusted for their size.[11] This is because smaller bodies lose heat more easily and because babies need to grow. In any case, babies need to eat relatively more. Thus, it wouldn't be so surprising to think that our brains are made to respond to the caloric composition of breastmilk with increased appetite. That only becomes a problem when it happens later in life where excess calories cannot help you grow much more than your waistline.

# 5

# What skydiving and salt have in common

It's a warm February morning in the Namibian city of Swakopmund. A tiny strip of asphalt is glittering in the desert sun as a rickety propeller plane speeds along and takes off. The plane is not in the best of conditions. It is missing the right door, and most of its windows are only held in place by duct tape. Inexplicably, the plane carries two young Danes, one of them dressed as a monkey and the other one dressed as a giraffe.

The monkey is me, and the giraffe is my friend Lars. During a trip across Africa, we've been egging each other on to go skydiving, and now we're getting a lesson in just how long half an hour can feel. It turns out that a ramshackle propeller plane takes an unnervingly long time to ascend.

Before the ascent we were fearless (hence the ridiculous costumes), but as Swakopmund disappears beneath us, the cabin goes quiet. At least the view is nice. Beneath us, the orange sand dunes of the Kalahari Desert dramatically meet the Atlantic Ocean along what is called the Skeleton Coast, or The Land God Made In Anger among the local San people.

Briefly, I forget the free fall awaiting us, but my daydreaming doesn't last. A powerful Arnold Schwarzenegger-accent easily cuts through the deafening sound of the engine.

'*Two minutz.*'

We review the safety protocols one last time.

Lars is closest to the door, looking a few kilometres straight down.

'*Rea-dy?*'

I don't get to hear his answer. In a split second, a large Danish teenager is sucked out of the missing plane door like a speck of dust being hoovered up by a vacuum.

Then it's my turn.

'*Rea-dy?*'

To this day, I'm not sure whether I even had the chance to respond.

*

So far we've met two members of the holy trinity of food superstimuli: fat and sugar. The last one we're missing is *salt*. One could argue that salt should actually take the throne when it comes to factory-produced food. You might be able to find this type of food with no added sugar. You can also find it without much fat. But it is difficult to find food made by food companies that doesn't have enough salt to de-ice a Scandinavian highway in winter.

Food manufacturers love salt because it is something of a miracle ingredient when you make food in a factory. It increases shelf-life, enhances the appearance and texture of food and, most importantly, it makes it taste better so people will buy more.

The reason we have a taste preference for salt dates all the way back to the first animals to leave the ocean and live on land. During our long evolutionary history submerged in seawater, salt was readily available. Animals made use of this resource by integrating salt into many important physiological functions such as nerve transmission and muscle contraction. However, on land, salt suddenly became much scarcer. At this point, the physiological processes that need salt were already so crucial that there was no way to re-engineer them to go without. Instead, land animals have to seek out salt in their diets to survive. To motivate this salt seeking, we developed taste buds for salty tastes that make us like foods containing salt. Herbivores especially like salt because plants contain less than meat does. This is why plant eaters from sheep to elephants will happily eat pure salt. We humans don't go quite that far, yet we are still instinctively drawn to salty tastes to keep our bodies functioning properly.

Does that mean food manufacturers are doing us a favour when they add tons of salt to all their products? Well, physiologically we need about half a gram of salt per day to survive. Our hunter-gatherer ancestors used to eat about 1.7 grams a day.[1] But today the average daily salt intake for a Westerner is around 9 grams.[2][3] And it would not be unheard of to find people eating twice that. In other words, the average Westerner eats an amount of salt that could cover the physiological needs of a large family. This enormous salt intake doesn't stem from people going mad with their salt shakers, though. For the most part the salt we add to our food ourselves is a minor part of our total salt intake. Instead, the majority of salt in the diet of modern people comes from processed foods. Once more, though, the sources are not always obvious.

Remember how added sugar is not only found in sweet foods and how bad we are at estimating the amount of fat in a dish. Added salt is also not just found in foods that taste salty. There can easily be as much salt in cheese, cold cuts and dressings as there is in crisps and French fries. Even foods that are only mildly processed often have added salt. Take chicken breast, for instance. In the supermarket, a pack of chicken breasts usually looks like the meat has simply been cut, packaged and shipped to the supermarket. But industrial chicken producers often use a technique called 'support' or 'plumping' in which they inject a saline solution into the meat. This makes the meat appear plumper and heavier. But it also enhances the taste because water evaporates when you cook the chicken, leaving the salt behind.

*

As you may have guessed, I managed to land on the ground in one piece after my skydive. Later, I got to talking with the instructor and asked him how a dive felt for him.

'I haven't had that feeling in my stomach for a while,' he said. 'For me, jumping out of a plane is as scary as going for a drive in my car.'

It might seem crazy to talk about a multi-kilometre free fall so casually, but it actually isn't strange or even unexpected. After all, we've learned that the body is an adaptation machine. And this is *especially* true of the brain, which adapts to whatever stimuli we are exposed to – even the wild ones like jumping out of a plane.

The first time you skydive your body will produce a rush of adrenaline as soon as it senses what is happening. There will also be a rush of endorphins in your brain that add to the euphoric feeling. But if you keep repeating the jump again and again – like my instructor – your body will slowly lower its response. Until one day, you're comparing skydiving to driving to work in the morning.

This phenomenon is called *desensitisation* and it's one of the most important concepts in this book. Basically, desensitisation is the loss of sensitivity to a stimulus we are repeatedly exposed to. This happens as the reward system in your brain slowly dials down its response when you repeat a rewarding activity. Desensitisation is why wealthy businesspeople routinely get bored of their mansions and supercars while other people might feel ecstatic about their first studio apartment or a new Opel Corsa. It's why thrill-seekers constantly chase bigger and more daring adventures. And it's a well-known phenomenon among drug addicts, too. They might start by taking a small amount of a given drug, which gives them a fantastic high. The dose keeps working a few times, but eventually desensitisation begins to set in. Then the drug taker has to either increase their dose or switch to something stronger to regain the feeling of that first high.

Desensitisation is also heavily involved in the obesity epidemic. Consider ice cream: it used to come in single flavours such as vanilla, chocolate and strawberry. But today you can buy ice cream such as those from Ben & Jerry's, which isn't just vanilla ice cream anymore, but vanilla ice cream mixed with caramel sauce, chocolate chips and cookie dough. Or with brownie pieces, peanut butter and marshmallows. These are superstimuli *bombs*. Vanilla ice cream is already a superstimulus with its high sugar and fat content. But it pales in comparison to the variety and sheer caloric density of something like a pint of Ben & Jerry's. So we can see the familiar pattern here. First, we are introduced to ice cream, which majorly triggers our reward systems. Maybe it becomes one of our favourite foods. But over time, we begin to get used to the superstimulus and start craving a little more.

How about adding some chocolate sauce on top?

Bingo.

The major reward is back.

But it doesn't last.

As we keep getting desensitised, we need to keep escalating. For now, that ends with a superstimuli bomb like Ben & Jerry's, but that's only until someone figures out how to create an even stronger superstimuli.

Now, not everyone will necessarily keep escalating. Just like not everyone who tries drugs develops a crippling addiction. But if you happen to be prone to desensitisation, you have a greater risk of escalating and therefore also a greater risk of becoming overweight in a world that is always ready to serve you the next superstimulus on the ladder.

If that sounds depressing, I agree with you. But fortunately, there is some good news as well. Our brains don't only get desensitised. The process can also go the other way in what is called *resensitisation*. As the name suggests, this is a process where we regain our sensitivity to a reward. And this is where salt re-enters the picture.

As you might recall, the average Westerner now eats enough salt to make a mermaid feel at home. This has happened because the large amounts of salt in ultra-processed foods desensitise us to salty flavours. You can imagine the whole thing as a negative loop, in which food manufacturers can gain an advantage over their competitors by using slightly more salt to make their products taste better. But as we are exposed to this saltiness, we gradually lose sensitivity, meaning food manufacturers have to keep upping the ante.

Fortunately, American scientists have discovered that we can regain our sensitivity to salt. In one study, the researchers recruited people who ate a normal American diet – which is even saltier than the British diet – and helped half of the study participants to lower their salt intake. Then the scientists followed people for five months, occasionally asking them along the way to taste soups and crackers with varying saltiness.

The research participants were not informed of the salt content of the food, but at first they preferred the saltiest products. Over time, though, the group that decreased their salt intake began to regain their sensitivity. The saltiest products became too salty for them, and they started to rate foods as saltier than previously.[4]

In other words, they had recovered their natural sensitivity and could enjoy food with less salt. At first the participants probably found their food flavourless. But the point is that it can be entirely 'free' to de-escalate in the long run. If you can survive the withdrawal period, you will be more sensitive to salty flavours, and you will enjoy your food as much as before.

★

It is not just our sensitivity to salt that can be regained after abstaining from superstimuli. We can resensitise ourselves to all kinds of taste experiences. Take sugar as another example. In a study reminiscent of the salt study above, scientists helped half of their study subjects to decrease their sugar intake. Over three months, the research subjects were asked to taste desserts with varying quantities of sugar and evaluate their sweetness. And the results were similar to the ones for salt. Those who cut down on sugar became increasingly sensitive to sweetness. They ended up rating the puddings as 40 per cent sweeter compared to subjects who had kept eating as previously.[5] This kind of experiment is easy to try on your own and I have done so many times. I can't recommend it socially and you don't need to completely abstain from sugar to be healthy. But it is interesting to feel how quickly resensitisation kicks in. After a few weeks without added sugar, you'll notice that fruits and berries taste *way* sweeter than before. And if you then return to sugary superstimuli like candy, it can feel cloyingly sweet and almost nauseating.

This is not only because the flavour is stronger, but also because of a principle called tolerance. Tolerance is another form of desensitisation – just to the *negative* effects of a stimulus. When we become 'tolerant', we lose our sensitivity to the harmful consequences of a substance. You might remember this from the first time you tried drinking alcohol. Back then, it probably didn't take much alcohol before you felt drunk. You probably also felt a little sick at the same time and might have revisited the drinks later in the evening – albeit not in the way you originally intended. But as you repeatedly drink alcohol, your body starts to build tolerance, and the unpleasant side effects of alcohol get less intense.

This is why alcoholics can drink the rest of us under the table and why you might notice your alcohol tolerance change depending on how much alcohol you drink in a given period.

Tolerance is also seen among smokers who manage to inhale toxic smoke without feeling ill. When people take their first drag of a cigarette, they usually cough themselves halfway to death. But just like with alcohol, repeated exposure builds tolerance. Importantly, this does *not* mean that the harmful effects of either alcohol or smoking disappear. In fact, our initial reactions are quite telling of how these substances affect the body. In the same way, the non-desensitised reaction to sugary candy is more telling than the one dulled by repeated exposure. So desensitisation and tolerance building make superstimuli bad twice over. When you eat food superstimuli, you gradually become desensitised to the reward stemming from the large amounts of sugar, fat and salt, meaning you will start to crave ever more. At the same time you build tolerance, losing your sensitivity to the negative health effects of each escalation.

**Superstimuli worldwide**

It's become standard to use the term 'Western diet' to refer to food superstimuli full of salt, sugar and fat. This is because burgers, French fries, pizza and so on originated in the West and have since conquered the world.

However, there is nothing specifically 'Western' about food superstimuli. The West just happened to be the first place in the world that was rich enough to develop the technology required to produce food superstimuli at scale. Over time, as the rest of the world has grown economically, other national cuisines have evolved in ways similar to ours.

In India, you can now get tons of salt, sugar and fat when you order butter chicken (containing butter, cream, flavour enhancers and artificial colours), with a side of naan (containing refined sugar and butter) and a mango lassi, a kind of Indian milkshake.

In China, you can order spring rolls that have been deep fried in tons of oil, followed by dishes like sweet and sour pork, which contain copious amounts of salt, sugar, oil and saturated fat. Then you can wash it all down with bubble tea, which has as much added sugar as soda.

Food superstimuli are simply such a good business case that they pop up everywhere, as soon as a country is rich and developed enough. And then it doesn't take long for obesity problems and lifestyle diseases to appear.

*

Desensitisation doesn't only happen over long periods of time. There is also a related phenomenon happening whenever you eat a meal. The first bite always tastes the best and then your reward diminishes bite by bite until finally you stop eating.

This kind of short-term desensitisation isn't permanent, of course. Next time you get hungry, the first bite will taste just as good again. But to understand the obesity epidemic, we also need to understand how short-term desensitisation works.

We have already learned that the feeling of fullness is in part physically conditioned by the stretch receptors in the stomach that tell us how full we are. But imagine that your friends have invited you over for dinner. They've made your favourite dish, and the first bite tastes fantastic. No matter how tasty the food is, however, there will inevitably come a point when you simply can't swallow another bite. Your stomach is full and says 'Stop!'

But then the host brings out the dessert . . .

Suddenly, your appetite returns and you have room for a few hundred more calories.

This phenomenon is called 'the Dessert Effect' or 'sensory specific satiety', if you want to be more scientific. Both concepts describe how the feeling of fullness isn't only physical. It is also partly mental and depends on what exactly you're eating. The physical satiety signals such as protein intake and stomach stretch don't disappear when dessert arrives. But the mental part of satiety does. This effect is also known as 'the Buffet Effect' because people tend to eat more before reaching satiety when they go to a buffet compared to if they are served à la carte instead.

Your brain seeks variety in foods due to evolution. Humans require many different nutrients and few foods can satisfy all our needs on their own. So we instinctively seek *variety* in what we eat.

As we learned from our Hadza friends, this desire made sense for our ancestors. Hunter-gatherers eat a monotonous diet by necessity. They are mainly limited to the food they can find within a few kilometres' radius. A good example is the !Kung tribe of Namibia. Just like the Hadza, they are generally in excellent physical condition. But I doubt you'd envy the diet that's responsible for that. Roughly half of *all* the calories consumed by the !Kung comes from one single source: mongongo nuts.[6]

I'm sure mongongo nuts taste great. Some say they're a little like cashews while others describe the taste as that of a nice cheese. No matter the taste, you would soon get bored if you had to get half of all your calories from mongongo nuts year after year. You would also risk developing deficiencies in whatever micronutrients the mongongo nuts lack. So it would make perfect sense for your brain to be designed to act more strongly if you encounter other foods and for you to instinctively want to look for variety.

Today we can see our desire for variation everywhere in our food landscape. In some cases, it is beneficial. Studies show that children eat more fruit and vegetables when they are introduced to a broad selection, even if the alternative is to give them their favourite fruit or vegetable only.[7]

More frequently, though, our love of variation is exploited to make us overeat the wrong things. Think of candy. It's no coincidence that it often comes in mixed bags. Even when it doesn't, there could still be superficial colour variation added such as in gummy bears or M&Ms.

Or think of Ben & Jerry's ice cream from earlier. The reason I call it a superstimulus bomb is because every ingredient – ice cream, cookie dough, chocolate etc. – is already a superstimulus on its own. But to escalate, several food superstimuli are combined into one, leveraging the draw of variety. so that the draw of variety is leveraged.

**Are food additives bad?**

Food manufacturers don't just optimise the taste of a food. They optimise everything from its sound – think of the crunch of crisps – to its texture, smell and appearance. To achieve these goals, the companies use various additives and these have often been blamed for causing health problems.

Of course, it is important to remember that food additives are not just random chemicals; they have to undergo extensive safety testing to make sure they don't cause health problems at the doses used. And as always, dosing is important. As we learned for artificial sweeteners, sometimes the negative headlines are due to studies on cell cultures or mice that use completely unnatural doses. Even water can be 'toxic' if you drink enough quickly.

So it's not that all additives are bad. Ironically, quite a few are also marketed as dietary supplements. You could probably picture someone you know vehemently denouncing 'E300' (also known as ascorbic acid) or 'E306' (also known as tocopherol) while promoting vitamins C and E. It just so happens that these are the same thing.

That said, even if many additives don't have an ill effect directly, it's important to remember why they are used: to make us eat more. Take flavour enhancers, which might not be toxic but which can harm health if they make us overeat and gain weight. Or take artificial colouring. These are added to food to manipulate us because humans are highly attracted to strong colours. Enhance the colours of your food product and more people will buy it. Again, not so different from the poor confused birds, huh?

# 6

# Health advice from the Amazon

By now, the causes of the obesity epidemic should be clear. We have gained so much weight because our food is *designed* to make us overeat. Food manufacturers do everything in their power to manipulate our appetite regulation. They play on our cravings for sugar, fat and salt. On our desire for variation. On our tendency to seek out ever stronger stimuli. *Everything* is optimised, down to the colour, texture and sound of the food. As a result, we are now faced with a global health crisis that has reached the most far-flung corners of the world. But what if you have managed to avoid becoming overweight despite this onslaught of food superstimuli? Should any of this still matter to you?

⋆

Deep in the Bolivian part of the Amazon rainforest lives a collection of small tribes called the Tsimané. These people call home one of the Earth's most remote places and they make a living growing simple crops, fishing and hunting in the surrounding jungle.

The close-to-nature lifestyle makes the Tsimané one of the few remaining examples of people living in more 'ancestral' or 'natural' ways than you and I. Just like our good friends the Hadza, that means the Amazon tribespeople are interesting for various researchers and receive frequent visits from academics around the world.

Scientists have studied everything from the social lives of the Tsimané to their sleep, diets and general health. In doing so, it quickly became apparent that it is the norm among members of the tribe to have chronic parasite infections of the gut. That's part and parcel of living in Earth's equivalent of a five-star resort for creepy-crawlies. However, the chronic infections mean that most Tsimané people have immune systems which are constantly active and fighting, giving them sky-high inflammation levels.

In the developed world, elevated inflammation markers are associated with many different diseases, among them what is called atherosclerosis. This is a kind of fatty plaque that can build up on the inside of the arteries as we age. You can imagine it as a pipe slowly clogging with grime. In a healthy artery, blood runs through easily, but the more plaque there is, the more difficult it becomes for the blood to flow. Eventually, the artery can be completely clogged, or some of the plaque can come loose. This causes the blood to clot, creating a clump that can obstruct blood flow and kill the tissue supplied by the blood vessel. This is especially dangerous if it happens in one of our two most important organs, the heart (causing heart attack) or the brain (causing ischemic stroke).

So if the Tsimané lived in the global West, their doctors would be urging them to resolve their inflammation issues. Of course, that type of privilege is usually not found in the rainforest. But the scientists doing blood work on the Tsimané got concerned and decided to study the cardiovascular health of the tribespeople just in case. Fortunately, it turned out there was nothing to worry about. The Tsimané didn't have clogged arteries. In fact, they had some of the healthiest hearts and cardiovascular systems ever seen. Even among the oldest members of the tribe, there was little to no atherosclerosis, meaning cardiovascular diseases were unheard of.[1]

To fully grasp how amazing this is, you should know that a third of *all* deaths in the entire world are due to cardiovascular diseases.[2] We are talking hundreds of millions of people, just in the time you've been alive. Heck, it is several hundred just since you started reading this chapter. The reason cardiovascular diseases hit so many is that atherosclerosis is really hard to avoid, at least if you live in the modern world. One of the first times we learned this was during the Korean War in the 1950s. When American doctors did autopsies on dead soldiers, they found atherosclerosis in 80 per cent of them. Most soldiers were young men in their early twenties and in good physical shape prior to passing. But the first signs of cardiovascular disease had already begun to appear while they were alive. Had fate been different, these young men would have returned home alive and well but with a ticking biological clock. Over the years, their atherosclerosis would have continued to develop, until one day it would have cost many of them their lives.

★

To understand why isolated tribespeople in the Amazon don't develop atherosclerosis we can follow the development of the disease.[3]

The first step is that something damages or irritates the inside of your arteries. This could be toxins from smoking or pollution that you have inhaled and taken up in your blood. It could also simply be high blood pressure straining the artery walls.

Once there is irritation or damage in the lining of the artery, particles from the blood called LDL (low-density lipo-protein) can get trapped in the area. You might know LDL as 'bad cholesterol'. That's because the normal job of these particles is to transport cholesterol around the bloodstream. So when the LDL particles get trapped in the lining of the artery, fatty streaks made of cholesterol start to appear. This cholesterol plaque alarms the body and it sends immune cells to the location tasked with cleaning up the mess. The immune cells get to work by eating up the cholesterol. But if there is too much, they can get overwhelmed and end up dying from exhaustion before completing the task. That means the plaque is now a mess of cholesterol and dead cells. As the body tries to remove the plaque, it summons ever more immune cells, making the plaque into an immune cell graveyard. All the while the cholesterol plaque just keeps growing larger. Finally, the body has an emergency solution: it encapsulates the plaque by enclosing it under a kind of lid so that at least it stops growing.

This saves the artery, but it isn't a great long-term solution. Now you have a narrowed artery containing a ticking time bomb. If the artery wall keeps being damaged, it is only a matter of time before the lid breaks open. Then the junk from underneath is released into the bloodstream, making the blood clot and that's when we reach the end of the line: blockage of the artery.

All those details might be a lot to take in. But here's what's important to know: atherosclerosis is fatty plaque that builds up in arteries. It starts when something damages or irritates the lining of the artery, after which cholesterol deposits start to build.

This disease progression gives us two possibilities to avoid the problem: avoid the initial damage or prevent the cholesterol deposits.

As mentioned, one of the most common causes of damage to the artery wall is having high blood pressure. This is also called having hypertension. Around a third of *all* people in the world are affected by hypertension but many don't know about it. It's not like there's steam coming out of their ears, and in fact, in most people there are no symptoms at all. Nevertheless, it is dangerous. Hypertension is actually the number one modifiable risk factor for premature death: it claims more lives each year than smoking, obesity and alcohol.[4]

If you want to avoid hypertension, there are several things you can modify but one of the most important ones is how much salt you eat. We already learned that salt makes food more rewarding and that can lead us to overeat and gain weight. However, weight gain is not the only health risk if you eat a lot of salt. An unnaturally high salt intake also increases the risk of getting kidney stones, of developing a host of autoimmune diseases and, most importantly, of getting high blood pressure.[5] [6] [7] The link to blood pressure has been demonstrated in studies where scientists make people adjust their salt intake. When people start eating more salt, their blood pressure rises. And when they cut down on salt, their blood pressure falls. This isn't only the case for humans, but has also been demonstrated in our close relatives, chimpanzees.[8] So while eating extra salt might seem harmless, it is actually a huge killer from a global perspective. Due to the connection with high blood pressure, our best estimates suggest that overconsumption of salt costs the lives of between one and five million people each year.[9]

Now, as we discussed, most of us don't have a higher salt intake than our ancestors because we go mad with the salt-shaker at dinner. We have higher salt intake because of food superstimuli. So given that the Tsimané and other tribespeople don't consume this type of food, they have low salt intakes and don't get high blood pressure.[10] In the West, we assume that hypertension comes with age. But it doesn't have to. Our blood pressure issues are a result of our lifestyles.

*

The other major villain in the story of atherosclerosis is the aforementioned LDL-cholesterol. We can see this in a genetic disease that cardiologists have specifically named to keep spelling bee participants up at night: 'familial hypercholesterolemia'. Let's just call it FH. People who suffer from FH have much higher LDL-cholesterol levels in their blood than normal. Some have so much that they develop yellow cholesterol deposits around their eyes, on their knuckles and on various tendons. The high LDL-cholesterol levels mean that people suffering from FH have a vastly increased risk of getting cardiovascular diseases. If they don't take cholesterol-lowering medications, half of the affected men will get some kind of cardiovascular disease by the time they are 50 years old. And a third of affected women will get one before they turn 60.

Fortunately, most of us do not have familial hypercholesterolemia. But the pattern is still the same. The more LDL-cholesterol you have in your blood, and the longer you are exposed to it, the larger your risk of developing fatty plaque in your blood vessels and dying of cardiovascular disease.[11] This is where we find the Tsimané's other big secret. The average tribesperson has much less LDL-cholesterol in their blood compared to the average Westerner.[12] Their LDL-cholesterol levels are simply too low to get atherosclerosis. We know this because Western people who, due to their lifestyle or genetics, have similarly low LDL-cholesterol levels also don't develop cardiovascular diseases. But even though the Tsimané are in the minority in a global perspective, they're actually not abnormal here. We are. The LDL-cholesterol levels of hunter-gatherers are similar to that of newborns and many animal species.[13] It's the natural amount. The only reason modernised people buck the trend is our diets. Sadly, this becomes evident even in the jungle. Because in the past decade, a few Tsimané people have acquired gas motors, which they use for boat trips to the various market towns of the rainforest where food superstimuli are for sale. The Tsimané are naturally just as vulnerable to food susperstimuli as we are. So, within the past few years, many of them have begun eating more ultra-processed food, and their LDL-cholesterol levels have begun to rise.[14]

In short, this happens because many food superstimuli are like a perfect storm for raising cholesterol levels. As we have discussed, sugar-sweetened beverages such as soda are known to do so in just a few weeks. Saturated fat in the diet also does.[15] And at the same time, food superstimuli are stripped of fibre, which naturally lowers cholesterol.

Dietary fibre is an umbrella term for various types of plant matter that we can't digest. You get it from pretty much all fruits and vegetables as well as from whole grains and beans. In the gut, fibre binds water which makes it take up space so you feel full. But because food manufacturers want to make sure you *don't* feel full, they remove fibre from their products. As a result, our fibre intake is markedly lower than our ancestors. The average Westerner eats something like 16 to 22 grams of fibre per day but traditional people like the Hadza consume 80 to 150 grams.[16] [17] [18] This aids the hunter-gatherers in staying lean by promoting fullness. And in turn, leanness is beneficial because excess body fat, especially fat around the organs, stimulates the liver to produce more LDL-cholesterol. Besides that, fibre also decreases cholesterol levels through a different mechanism.[19] [20] [21] The mechanics of this one are a little messy, but it goes like this:

Your intestines use molecules called bile salts to absorb fat from your food. Bile salts are made in the liver and stored in the gallbladder. When you have eaten, they are excreted into the intestines. Here, the bile salts bind to fat molecules and make them easier to break down and absorb. As we know, the body is primed for conserving energy and resources. So once the job is done, the body reabsorbs the bile salts to reuse them. But that is not always possible if you eat a lot of fibre. If there is fibre in your intestines, some of the bile salts get trapped in the tangles and are *not* reabsorbed Instead, they get excreted from the body with the rest of the waste from digestion. This might not sound like a good thing, but bile salts are made from cholesterol, so when you lose them, you need to make new ones. Your liver does so by absorbing LDL-particles from your blood and using the cholesterol as starting material.

**Eggs are okay**

In the 1990s, an American doctor stumbled on a man with dementia who had a peculiar eating habit. Every day for years, this man had been consuming pretty much nothing else but 25 soft-boiled eggs. Given that eggs contain cholesterol, that is an enormous cholesterol intake. Did it cause his arteries to clog up? No. Even though the man was eating large amounts of cholesterol, his blood work was entirely normal.[22] This is because, for most people, cholesterol in the diet only minimally impacts the amount of cholesterol in the blood.[23] Our bodies make their own cholesterol. So generally, if you eat more cholesterol, your body will turn down production and absorption from the diet. Vice versa – if you eat very little cholesterol, it will turn up absorption and increase local production. So, in other words, even though eggs contain lots of cholesterol, they don't generally increase LDL-cholesterol much. There are some people, called 'hyper-responders', for whom this is not the case. But for the majority of people, eggs are healthy.

*

Maybe you're starting to think that it is necessary to live like a hunter-gatherer if you want to be healthy. Certainly, there's no doubt that the various tribespeople we have met so far are very metabolically healthy and have low risks of diseases such as diabetes, heart attacks and strokes. But perhaps this knowledge conflicts with what you have previously heard. Don't people usually say that the average Stone Age lifespan was under 40? That doesn't really align with being ultra healthy.

It's true that the average lifespan of our ancestors was low. And the same goes for contemporary hunter-gatherer populations. The Hadza life expectancy at birth, for example, is only 34 years.[24] But this doesn't mean that most Hadza people die from age-related causes in their thirties. No, the explanation is that most hunter-gatherer societies have a massive rate of infant mortality. Four out of ten Hadza die before they reach the age of 15. This pulls the average down. In a group of four, where two die at the age of 80 and the two others die as infants, the average life expectancy is 40.

But if you take infant mortality out of the equation, the life expectancy of hunter-gather populations resembles our own much more. Among the Hadza who live past 15 years of age, the most common age of death is 76, and it is 78 for the Tsimané.[25] That's remarkable considering their lack of access to modern medicine or technological assistance.

The most prevalent cause of death among the Hadza – for both adults and children – is infectious disease. Seventy per cent of all deaths among contemporary hunter-gatherer populations are caused by viruses and bacteria. So you might say that the healthiest person of all would be a combination: a Hadza with modern health insurance.

Of course, it's not realistic for you and me to live like the Hadza. Your neighbours might look at you strangely if you start hunting the local squirrels for food. However, there *is* a food movement that attempts to combine modern life with the best parts of the hunter-gatherer existence. This is what's called the 'Paleo diet'.

This diet excludes all foods that our Stone Age ancestors wouldn't have had access to: starches like bread and rice, various kinds of beans, dairy products and practically every ultra-processed food. Instead, the diet is based on meat, fish, fruit and vegetables. There's no doubt that this diet is healthier than the average Western diet. But in reality even the 'Paleo diet' must admit defeat. Because, despite the name, the Paleo diet is nothing like what Palaeolithic humans actually ate. It turns out, for instance, that grains aren't as recent an invention as we thought. Archaeologists have found evidence of bread from thousands of years before we started cultivating land. This suggests that we ate various kinds of grass – the ancestors of grains – for even longer.[26 27 28]

On the other hand, many foods that are encouraged in the Paleo diet were unknown to our ancestors. Tomatoes, chilli peppers and potatoes are all encouraged on the Paleo diet as they are vegetables. But recall that they are all new foods unless you are of Native American descent. These foods didn't leave the American continent until a few hundred years ago and thus Europeans have been eating 'banned' foods such as wheat and dairy far longer than tomatoes, chilli peppers and potatoes.

Furthermore, we have learned that even the foods that we *did* actually eat in the past have been heavily modified since then. Fruits are much bigger and sweeter and meat is fattier. Even our own bodies are not the same anymore. Populations like Europeans and Asians – who have been farming for a long time – have genetically changed since the Stone Age to become better at digesting starches such as the ones in bread and rice. And in Europe, parts of India and Africa, we see genetic adaptations to eating dairy products.

All of this is fine, though, because you don't have to go all the way back to the Palaeolithic for inspiration. Many of today's primitive farming communities are just as healthy as hunter-gatherers. Our friends from the Amazon, the Tsimané, for example, do hunt in the rainforest, but they get most of their calories from crops, including plantains, rice, cassava root and corn.

So in our quest for a healthier life, we don't need to roll back time thousands of years. Much less will do.

Let's dive in.

# 7

# The spud diet

'Make your food boring and your life interesting,' was Andrew Taylor's motto when he set out to lose weight.[1]

Taylor is from Melbourne, Australia, and he used to weigh 151 kilograms. Like many people, he knew he was eating all the wrong things but he just couldn't help himself. As he considered his relationship to food, Taylor came to the conclusion that it resembled an addiction. He didn't just like ice cream and junk food. He *craved* it. So in an effort to fight back, he decided to turn to the oldest of addiction cures: going cold turkey.

The problem with this treatment, of course, is that going cold turkey on food is not exactly a long-term solution. Eventually you will have to eat *something*. To avoid starving himself, Taylor decided he could mimic going cold turkey not by eliminating food entirely but by focusing on eliminating his pleasure-seeking way of eating. He just needed to find the most boring diet possible and stick with it.

I think we can both agree there are a few worthy candidates here, but Taylor ultimately decided on a diet consisting solely of potatoes. Not too far off, if you ask me. At first, Taylor only intended to follow his strict potato diet for a few weeks but as the kilos started rattling off, he decided to keep going. You can't get all the nutrients your body needs from potatoes alone, though. So, along the way, Taylor made a few adjustments to keep himself healthy. There's not enough vitamin A in potatoes, but Taylor overcame that by including sweet potatoes in his diet. There's also no salt or vitamin B12 in potatoes, so Taylor salted his potatoes and started taking a multivitamin.

Besides those adjustments, though, Andrew Taylor lived off nothing but potatoes. He let himself eat as soon as he felt hungry and he ate as many potatoes as he wanted. In other words: hunger was never a problem. That's a big plus compared to most weight loss diets. But although Taylor was never hungry, he ended up in a caloric deficit. He simply couldn't eat enough potatoes to maintain his weight. The experiment went on for a whole year, and by the time Taylor was done, he had lost 53 kilos. At the same time he had normalised his formerly high cholesterol levels and blood pressure.

Since his spud days, Taylor has transitioned to a more varied diet. As he puts it, the point of the experiment was never to find the optimal diet to live on in the long term. It was simply an effort to reset his unhealthy relationship to food. And it worked. Taylor has managed to keep the weight off since, which you might remember is rare.

His secret? Still potatoes. When I spoke with him, I asked him about his latest meals and the answer was: a large portion of plain potatoes, just like the old days.

Now, obviously, a diet as monotonous as the potato diet isn't healthy long term. Yet, I know from experience that no matter how much I follow up a story like this by saying 'please don't try this at home' and 'this is a bad idea', a few people are going to read this and immediately think 'that sounds fun, let's try'. So I figured that I'd have to try the potato diet myself just to see what kind of torture I would be putting my poor readers through. Okay, maybe also a little bit due to the wackiness of the idea. In any case, I convinced my brother to participate, so one Sunday we ate our last proper meal and then started preparing various kinds of potatoes for the following five days.

Lesson one: people in the lunchroom at work look at you weird when your plate looks like that of a nineteenth-century Irish peasant.

Lesson two: it is absolutely impossible to get enough calories when eating nothing but potatoes. At least I couldn't.

You see, potatoes are actually one of the most filling foods known to mankind. I could attest to that much while trying to force-feed myself on plain baking potato number three. But it is also backed up by scientific studies. A group of Australian scientists have created what they call the 'satiety index', which is an overview of how filling given foods are per calorie.[2] In other words, it tells you how full you'll feel from eating 100 calories of apples compared to 100 calories of oranges.

To construct their satiety index, the Australian researchers recruited a bunch of students who were served 240 calories of a single food source. Then the students were asked to rate how hungry they felt every 15 minutes over the next two hours. And finally, they were given access to a buffet, where the scientists secretly recorded how much each student ate.

You might think that the subjects would feel equally full, no matter what they were served by the scientists. After all, they were all getting the same number of calories. However, the level of hunger varied widely. Some of the participants felt stuffed when the buffet arrived, while others dug in.

The least filling foods won't surprise you: these included cookies, candy bars and other food superstimuli that, as we know, are designed not to be filling. Further up the satiety index ranked various whole foods such as beans, meat and fruit. But even among the whole foods, one food was vastly more filling than all the others. This food was – drumroll – boiled potatoes. And it wasn't even close. Other sources of starch like bread, rice and pasta were much less filling.

When I first read this study, I wondered about the discrepancy. Why would potatoes be so much more filling than other sources of starch? After my experiment with the potato diet I'm not wondering anymore. I didn't feel hungry a single time in my five days eating nothing but potatoes. That's not to say it was a particularly pleasant experience. Quite the contrary, in fact. I wouldn't recommend it. But I wasn't hungry. Why not is uncertain but here are a few guesses.

*First*, potatoes contain fibre, which as we know is satiating. However, many other whole foods on the list also contain fibre. Often even more than potatoes, so that can't be the sole reason. Moreover, much of the fibre in potatoes is in the peel, and when doing the potato diet it is important to peel the potatoes because of compounds in the skins that are normally no problem but which can become problematic when one eats nothing else.

*Second*, potatoes contain a protein called proteinase inhibitor II. In experiments on mice, it's been shown that this protein reduces appetite and the same could be true in humans.[3] [4]

*Third*, there is a hedonic aspect. If you ask me, boiled potatoes are one of the most boring foods in existence. You might even say that boiled potatoes are an anti-superstimulus. Flavourless, fibrous and monotonous.

*Fourth*, the energy density of potatoes is surprisingly low. This means that potatoes contain very few calories relative to their size. A boiled potato might seem heavy, but much of it is water and it doesn't have a lot of calories: 100 grams of potatoes come in at 77 calories. Thus, a kilogram has 770 calories and you would have to eat several kilograms just to get enough calories even to reach your recommended daily intake and maintain your body weight. That's no easy task.

In the Australian experiment, the low calorie density of potatoes meant the participants had to eat over 300 grams of boiled potatoes to get as many calories as the people who ate a single Mars bar. So, it isn't shocking that the potato eaters felt fuller; their bellies simply had more volume in them. Overall, this is a good rule to remember. Foods with a low calorie density are more satiating than foods with a high calorie density.[5] So a quick rule of thumb is to look for foods that produce as much satiety as possible for the as few calories as possible.

Most unprocessed foods, besides nuts, do well in this aspect. For example, 100 grams of strawberries only have a paltry 32 calories. So if that's your snack, you can eat *750 grams* (40–60 strawberries) to equal the calories in a single Mars bar.

★

By now you have probably realised that the diet world is a bit of a jungle. There are practically online wars fighting over which one is superior. Some swear by a low-carb diet, in which you don't eat pasta, bread, rice and so on. Others swear by a low-fat diet, which is the complete opposite. Some people lose weight by becoming vegetarian and avoiding meat. Others lose weight by eating *more* meat. And the list goes on from the

Mediterranean diet to the aforementioned potato diet and the even wackier gummy bear or cotton ball diets.

However, in reality this space is actually much less complicated than it seems. Let me fill you in on a little secret: most of these diets work – just not why people believe they do. People want to blame carbs, sugar, fats and so on but in reality the reason restrictive diets work is that they make eating more boring.

Take it from someone who's tried them all. Eating is less rewarding when you cut out an entire food group. No matter how exciting diet book authors claim that low-carb or low-fat food can be, bread and pasta taste good. Meat and fish taste good. Bacon and cheese taste good. So if you cut certain foods out of your diet, eating will simply be less exciting. Moreover, there will be less variation in your diet and as we've learned, variation in itself makes us eat more.

We can see this phenomenon of diminishing rewards in studies of vegans and vegetarians. The closer people get to pure veganism, the fewer calories they eat.

Omnivores eat the most calories.

Next come pescatarians, who cut out meat but still eat fish, shellfish, eggs and dairy products.

Then it's vegetarians who also cut out fish and shellfish but still eat eggs and dairy.

And finally – the ones who eat the fewest calories – are the vegans who don't eat any animal products at all.[6]

So the rule is that the more foods you eliminate – the more restrictive your diet is – the fewer calories you'll typically eat. This is also reflected in the average BMIs seen in these diet groups. Vegans have the lowest BMIs, followed by vegetarians, pescatarians and omnivores.[7] [8]

The potato diet simply takes this principle to the extreme by restricting veganism even further. If a group of people actually lived full time on the potato diet, their caloric intake and BMIs would undoubtedly be the lowest of all.

So to sum it up: restrictive diets work because you eat less when your food is less rewarding. The same phenomenon can be seen in our old friends the Hadza, as well as in other hunter-gatherer populations. The researchers studying these populations always have a lot of nice things to say about the tribespeople: they are warm and welcoming, take great care of their guests and are fun and enjoyable company. However, some researchers will sheepishly admit that if they *had* to complain about something, it would be the food. It's really bland.

As the anthropologist Herman Pontzer, whom we met earlier, writes in his book *Burn: The Misunderstood Science of Metabolism*:

> Hadza food isn't very exciting. Aside from the honey and some of the tangy fruits, it's all quite bland. Spices are unheard of, other than the occasional sparing dash of salt. Nearly all the food is served on its own, either raw, roasted, or boiled. It's not what most Westerners would describe as tasty or even appealing. No food is too bloody, too old, or too unsightly. If you've ever opened your grill the day after a big barbecue to find a cold, forgotten drumstick and a lonely potato blackened on the grate, you've encountered Hadza cuisine.[9]

Pontzer's description gives us a sense of how our ancestors ate for millions of years. They didn't conjure up exciting dishes with tons of ingredients or have access to much variety. Instead, they ate the same foods over and over, pretty much always prepared the same way. And because obtaining food required such great effort, they were willing to eat food that we wouldn't touch in the modern world. *Bon appétit.*

To be fair, though, it is not clear that hunter-gatherers *themselves* find their food unexciting. Are they bored eating game meat and fibrous tubers over and over? Doesn't seem like it. Most Hadza mention food as one of the top things that make them happy in life and they celebrate good meals. This makes perfect sense when you remember that our brains are adaptation machines. If you have been exposed to food super-stimuli all your life, you become desensitised and 'natural' food is unappealing. But if this type of food is all you know, your preferences form around it. That's clever because the less desensitised you are, the less sugar, fat, salt and so on you need to feel good about your diet. As we discussed, fortunately we can resensitise, at least to a certain extent. You and I will likely never be as excited about a sour fruit as a Hadza. But it is still mighty helpful to aim at some resensitisation when trying to lose weight or get healthier. And while we have established that we can never copy the Hadza accurately, we can still gain from reducing the variety of our diets a little from time to time.

**Why is plane food so boring?**

Colorado is the highest located state in the USA with an average elevation of 2,100 metres above sea level. It is also the leanest state in the Union and surprisingly the two might be connected. At least, it seems like a general phenomenon that altitude is somewhat protective against becoming obese.[10] [11] For instance, a study that followed US military personnel found that those stationed at higher altitude tended to gain less weight.[12] The reason why might be that food is less appealing when air pressure is lower. If you've ever tried eating on a plane, you will have noticed that this is not exactly the height of culinary experiences. And in fact, a study conducted by the airline Lufthansa found that taste bud sensitivity to sweet and salty foods is up to 30 per cent lower while in the air.[13] Additionally, the low air pressure and humidity reduce our sense of smell.

So maybe mountain dwellers have an easier time staying lean because their surroundings make them less vulnerable to food superstimuli manipulation.

# 8

# From dynamite to diet pills

A few weeks into World War I, German troops advanced into northern France with the intention to reach and capture Paris. The very fate of the French Republic was at stake and the French army met the Germans with fierce resistance, fighting some of the bloodiest battles Europe had ever seen.

Behind the scenes, workers at French bomb factories were toiling around the clock to produce explosives for the effort. Safety standards at these factories had already been lax before the war, but with the survival of the nation at stake, they dropped even further. All that mattered was the quick delivery of explosives to the front lines. Thus, the factory workers were often in close contact with the chemicals used, inhaling and touching them, and over time many began to report a battery of unusual health symptoms.

They were sweating profusely – especially at night – felt utterly exhausted and were rapidly losing weight. For some, the symptoms grew so bad that it cost them their lives. During the war, focus was elsewhere, but upon the conclusion of fighting, scientists decided to revisit the issue. What had caused the poor bomb factory workers their misery? The researchers began studying the various chemicals at the factories to explore their effects on laboratory animals and humans. After years of investigation, their conclusion was that the symptoms stemmed from a chemical called dinitrophenol, or DNP.[1]

DNP is used in explosives because it is highly combustible. If you compact it under pressure and use the right detonator, it can be made into a formidable bomb. However, it turns out DNP has another effect if it is inhaled or ingested by biological organisms. The scientists showed that DNP affects special structures in our cells called mitochondria. If you immediately thought 'the powerhouse of the cell!' you might want to decrease your screen time. But you'd also be right. Mitochondria are so-called called organelles, or cellular organs. They gained their 'powerhouse' nickname because they are responsible for the final steps in extracting energy from our food. This process ultimately ends in the production of molecules called adenosine triphosphate, or ATP. Imagine it like this: first, you eat some food. This food is digested in your stomach and intestines, after which the nutrients are absorbed into the body and end up in the bloodstream. From here, the nutrients are taken up by cells which begin extracting energy from them. In most cases, the first couple of steps occur in the main compartment of the cell, which is called the cytoplasm. But the majority of energy extraction happens inside the mitochondria. The resulting energy is not used right away but stored in molecules called ATP, which you can imagine as little batteries. By making ATP, the cell can save the energy for later and use it for various energy-intensive tasks. For instance, a muscle cell could use ATP to power contraction or an immune cell could use ATP to power its fight against an infection.

Okay, that was a lot of molecular biology. The main take-home point here is that your mitochondria – cellular powerhouses

– extract energy from your food which they 'save' for later by making ATP. This is the fundamental phenomenon keeping you alive and powering all your various bodily functions. However, DNP interferes with the mitochondria. It makes the energy extracted from your food get burned off as heat instead of being saved as ATP. Picture it as putting your car in neutral and pressing down the accelerator. If you do so, you won't move forward. The engine will just get hot. So when the French workers were inhaling DNP from the explosives, they burned energy as heat instead of making ATP. That is why they got hot and were sweating so much. In turn, they had low ATP levels and thus lacked energy for cellular tasks. That explains the feelings of extreme fatigue. When the bodies of the bomb workers tried to fix the energy shortage by burning body fat, this energy was *also* just turned into heat, making the sweating even worse and explaining the rapid weight loss.

Now, none of these symptoms sound particularly pleasant. But can you guess what happened next? Some of the bomb workers had burned 50 per cent more calories than normal while being exposed to DNP. So soon thrifty entrepreneurs were marketing the compound as a way to effortlessly 'melt away fat'. Within a few years, more than 100,000 people had started taking DNP hoping to slim down. It had become the first 'diet pill'. Of course, that moniker didn't make DNP any safer. People were still dying in macabre ways. One doctor recalled admitting a DNP user whose body temperature reached 45°C before he died. Doctors also started discovering new side effects that had not originally been described. Some people developed rashes. Others awoke with blurred vision or suddenly went blind.[2] [3] Not surprisingly, doctors grew increasingly concerned and in the end DNP was declared illegal. In fact, this happened as soon as health authorities were granted legal power to ban substances. The ban caused DNP use to drop drastically but it has never fully disappeared from the black market. Among others, it has often been used by bodybuilders to lose weight before competitions, and just like the old days it continues occasionally to claim lives.

Despite all these grim fates, the story of DNP did not only lead to misery, though. In fact, the discovery kickstarted the hunt for medical obesity treatments. After all, if you disregard the horrible side effects, the fundamental principle is promising. If we could develop drugs to safely increase our metabolism, that would be a simple and satisfying solution: eat whatever you want, and then increase your energy expenditure with a pill so you don't gain weight.

This idea, unfortunately, has turned out to be more complicated than one might have hoped. Obesity researchers have discovered several drugs besides DNP that increase metabolism but they all seem to come with unfortunate side effects. Most often, these drugs end up causing heart problems. So instead, researchers began focusing on the other side of the equation: making drugs to lower appetite so people eat less.

★

One of the most important inspirations for appetite-suppressing medicine came from cannabis smokers. Cannabis tends to induce something smokers call 'the munchies'. This is a sort of eating frenzy characterised by cravings for candy and snacks. In this sense, cannabis has the completely opposite effect of what obesity researchers were looking for. An *increased* appetite. This got researchers thinking: what if you could make an *anti*-cannabis? Something to give you the *anti*-munchies?

One French pharmaceutical company started looking for a way to do this and found a compound that affects the brain in an opposite way to cannabis. They named this compound rimonabant, and in clinical trials it really did produce something like anti-munchies. The study participants had less appetite than usual, ate less and ended up losing weight.[4] All signs pointed to this *anti*-cannabis becoming the first proper pharmaceutical treatment for obesity. It even managed to be approved in the EU and thousands of people received prescriptions from their doctors. But in the final clinical study in the United States, there were problems. Maybe they seem obvious in hindsight.

Cannabis is primarily used because it produces feelings of well-being. At least temporarily. So an *anti*-cannabis might be expected to impact your mood in a negative direction. In the American study, scientists found that up to 10 per cent of rimonabant users developed depression. And the trial had to be stopped after two participants committed suicide.[5] Marketing authorisation of the medication was subsequently suspended in the European Union.[6 7]

This failure was a demoralising blow for obesity research. Yet again: so close, yet so far. It looked like we would never be able to find a medical treatment for obesity. DNP made people blind or nearly boil to death from the inside. Rimonabant made people depressed. And every other attempt was failing as well.

The whole ordeal is perfectly exemplified by a weight loss product called Olestra. This was an experimental food ingredient that researchers hoped would be something like artificial sweeteners but for fat instead of sugar. That is, an ingredient that would trick us into thinking we were eating fat but that contained a lot fewer calories. Olestra performed well in clinical trials where people successfully cut down their calorie intake when replacing fats with Olestra. But when the products hit the market and people started using them at home, some unfortunate side effects appeared. It turns out that people behave differently in a clinical setting compared with how they behave at home on the couch. In a trial while being followed by researchers, study participants might eat an Olestra cookie or two and then stop. But at home away from judgemental eyes, the story is different. It is not particularly hard to make it through an entire box of cookies instead of just a few. And when people did so with products containing Olestra, some began feeling bad. The symptoms ranged from stomach pains to explosive diarrhoea. In the end, most food producers chose to withdraw their Olestra products.

As an obesity researcher back then, you might have felt like the punchline to some kind of cosmic joke. Obesity rates were exploding and all the scientific effort managed to do about it was give people heart problems and diarrhoea.

*

Despite the many setbacks, obesity medicine was not destined to fail forever. In 2021, Danish pharmaceutical company Novo Nordisk announced the result of a clinical trial for an obesity medicine called semaglutide. It is also known by its trade names

Wegovy and Ozempic. In the Novo Nordisk trial, participants were advised on how to use diet and exercise to lose weight. And then two thirds were given the drug while the rest were given a placebo. At the beginning of the trial, the average participant weighed 105 kilograms. After a year and four months, those only receiving guidance on diet and exercise had lost 2.6 kilograms. Those getting Ozempic on top, though, had lost 15 kilograms, or 15 per cent of their original body weight.[8] [9] That kind of slimming was unprecedented for practically any intervention. At the same time, the side effects had been limited to stuff like nausea and constipation and were usually transient. When this news hit the world, it started an absolute frenzy. The Novo Nordisk stock rose so much that it became the most valuable company in Europe. At one point the valuation was equivalent to the entire Danish GDP. Even many experienced investors hadn't expected such a response and for most normal people Ozempic practically appeared out of thin air. However, the whole saga actually began several decades earlier.

In the late 1980s, Danish researcher Jens Juul Holst and his colleagues were doing a series of experiments on isolated pig organs to study a hormone called GLP-1.[10] That might sound like one of those passwords the IT department forces you to change every 90 days. But it is actually a hormone that is released by certain cells in the intestine after you eat. Holst and his colleagues showed that GLP-1 in turn stimulates the release of another hormone called insulin. This was a big deal because insulin was the most valuable hormone in the world at the time (ironically, now it is GLP-1).

Insulin is important because it is used to treat people suffering from type 2 diabetes. It works like this: at any given time you have about a teaspoon-worth of sugar in your blood which your cells can tap into for energy. However, after meals containing carbohydrates such as bread, rice or potatoes, sugar comes rushing in as you digest. That sounds scary but it isn't. The rise in blood sugar causes your body to release insulin, which makes your muscles, fat and liver take up the excess sugar and either burn it for energy or store it for later. This is totally normal physiology and happens every single time you eat. Problems only arise if this mechanism goes wrong. If you have type 2 diabetes, you gradually lose your response to insulin and after a while production can also go down. The end result is that you have heightened blood sugar and this is bad news. Too much sugar in the blood is toxic to blood vessels, and in the late stages of type 2 diabetes, damage can be so extensive that patients go blind or have to get amputations. Thus it is paramount to help type 2 diabetics improve their blood sugar control and GLP-1 looked to be a promising new way to do this.

Before long, it became evident that GLP-1 did more to the body than just regulate blood sugar, though. Scientists described how rodents infused with GLP-1 for extended periods began to lose weight.[11] In hindsight, this is even more exciting than the blood sugar effects. But remember the track record of weight loss drugs. For most pharmaceutical companies, trying to treat obesity sounded like a foolproof way to set cash on fire. So they mostly focused on trying to make GLP-1 into diabetes medication first.

To do so, researchers faced one problem in particular: when you inject people with GLP-1, it disappears in minutes.[12] This is because GLP-1 is small enough that it is filtered out in the kidneys and because an enzyme in blood vessels actively destroys it. You can think of these enzymes like little scissors that won't cut up just any old molecule but will specifically cut up a select few, including GLP-1. In both cases, you cannot really treat anyone if the molecule you use disappears immediately, so pharmaceutical companies began experimenting.

The first solution came from a small American company called Amylin Pharmaceuticals. Scientists here found the answer in the most obvious of places: the spit of a venomous desert lizard called the Gila monster. This lizard is native to the Western United States and for some reason it produces a peptide in its saliva (not venom) which is similar to GLP-1.[13] To this day, no one really knows why. However, the cool thing is that this peptide is close enough to GLP-1 that it works the same if you inject it into humans – but different enough that it does not get cut up in the blood vessels. So, the Gila monster peptide ended up being developed into a new diabetes medication called exenatide. It was far from perfect, though, and it never became a blockbuster drug. Even though the drug didn't get cut up or filtered out by the kidneys right away, it disappeared within a few hours, so something better was needed.

*

In May 1994 a young researcher came storming into her boss's office with a newly published research paper in hand. Lotte Bjerre Knudsen was a chemist employed at Novo Nordisk. The paper in her hand was co-authored by the aforementioned Jens Juul Holst and showed how injections of GLP-1 helped type 2 diabetics normalise their blood sugar. Novo Nordisk, specialising in diabetes medication, was obviously interested and so the journey towards Ozempic began.[14] Bjerre Knudsen was tasked with finding a way to make the GLP-1 hormone survive in the body for 24 hours. She and her colleagues didn't do so by turning to the venomous Gila monster but by combining GLP-1 with a fat molecule. This is clever because of the way fats are transported around the body. You might remember from chemistry class that fats are not water soluble. For instance, if you pour oil into water, it won't mix. Rather, the oil forms a separate layer. This phenomenon causes us problems when we want to transport fats around our body, because our blood is mostly water so we cannot transport the fats as they are. Instead, we use various other molecules to help. One of them is an abundant blood protein called albumin. This protein can bind to fat molecules and drag them along, so you can picture it kind of like a little bus that picks up and drops off fats and other molecules. By combining GLP-1 with a fat molecule, Bjerre Knudsen made sure it would attach to the little buses, albumin. In turn, that protected the GLP-1 molecule from being cut, made sure it wasn't filtered out in the kidneys and helped it stick around for days. Novo Nordisk named this new drug liraglutide and it ended up being approved as medicine, first for the treatment of diabetes and then for weight loss. In the final clinical trial, participants weighed 106 kilograms and lost 5.6 kilograms more than those who simply ate better and exercised.[15]

To make Ozempic, Novo Nordisk changed liraglutide by using another type of fat molecule while slightly changing the GLP-1 molecule itself. This made the drug more potent by sticking around even longer and as a result people lost 15 per cent of their body weight versus the 5 per cent for liraglutide. Not long after, American pharmaceutical company Eli Lilly introduced a GLP-1-based drug, tirzepatide, that combines this effect with mimicking another gut hormone that also suppresses appetite. As a result, this drug made participants lose 21 per cent of their body weight in the final clinical.[16] Novo Nordisk answered with its own double drug that slightly beat *that*. And at this point there is pretty much not a single pharmaceutical company that doesn't have a new competitor on the way.

One question remains, though. How do these drugs even work? Why do they make people lose weight? First it was thought that GLP-1 makes people slimmer by slowing down the movement of food through the intestines. After all, it makes some people constipated. The thinking was that people felt full for longer due to the higher amount of food in the intestines at any given time. However, studies in laboratory animals showed long ago that the gut effect is not needed for weight loss. Instead, these weight loss drugs work by affecting *the brain*.[17] Specifically, they affect areas of the brain that have to do with appetite regulation and reward. So the mechanism is that they make people feel less desire to eat, lower their calorie intake and thus slim down. The effect seems to be particularly pronounced for 'salty snacks and sweets' i.e. food superstimuli.[18] And in fact, it is already being felt in the real world. The CEO of Walmart reports that the company sees falling sales for food superstimuli such as candy and crisps. And data from the US shows obesity rates actually fell a little bit recently for the first time since record-keeping began.[19]

**More calories could also be a good thing**

In this book we focus on the fact that food superstimuli make us eat too many calories, which has all sorts of negative downstream effects. However, we could also look at this problem from another angle. You could say that our bodies are really shooting themselves in the foot here. Storing excess calories as fat made perfect sense in our evolutionary past. We have big brains that require a lot of energy, a requirement which is difficult to ignore. So when you live in an environment where food is hard to come by, it is a good strategy to save some of the energy from your food for later. Now, though, this storing strategy just makes us sick. In a perfect world, our bodies would use the excess energy we have available productively instead. There are endless possibilities here: we could have boundless energy for activity, have even larger brains, have stronger bones or all be *buff*. Humans are weak compared to other animals, likely because our brains cost so much energy that we had to save energy by lowering muscle mass. But in the modern world, we could have both. I'm sure you wouldn't mind having to eat some extra ice cream in return for looking better at the beach. For now, though, this is mostly a pipedream and ever-better appetite suppressors is the name of the game. Currently we just can't safely push our bodies towards higher and more productive energy expenditures but let's see what the future brings.

# Part 2

## SEX AND DRUG SUPERSTIMULI

# 9

# Sex and the bottle's spell

Western Australia is known for its so-called bush, an iconic landscape of red brown earth and scattered shrubs which looks iconic on a postcard. For plants and animals, though, the bush is not exactly a paradise. It is bone dry, roasting hot, and the soil is some of the most nutrient-poor soils on Earth. As a result, Western Australia isn't exactly bursting with wildlife. Even humans are sparse on the ground. Eighty per cent of Australia's population lives along the more fertile east coast, and the only large city in Western Australia, Perth, is one of the world's most isolated major cities.

Ironically, though, the absence of humans doesn't mean the animals of the bush are free of human-made problems. Take the jewel beetle. You'd probably have to be an entomologist to name this insect after expensive gemstones. Most people would say the jewel beetle looks more like a cockroach with its brown oblong body and black head. But though the jewel beetle won't win any beauty pageants, it is an excellent example of adaptation. Against the red-brown backdrop of Western Australia, the jewel beetle is well camouflaged and it spends most of its time undisturbed, feasting on various plants in the dry landscape.

However, once a year, this monotonous lifestyle is interrupted when mating season arrives. Then the male jewel beetles take flight in search of females to mate with. The females can't fly, so the lovestruck males buzz around the bush while frantically scanning the ground for a brown mate. This is where the human-made problem begins. Thirsty Australians sometimes leave their empty beer bottles behind and this becomes love at first sight for the poor jewel beetles. The empty beer bottles are large, shiny and brown, just like a female jewel beetle. In fact, the bottles are larger and shinier than the female beetles could ever hope to be. They're a sexual superstimuli.

So, if a male jewel beetle spots an empty beer bottle, he buzzes right over and fervently tries to mate with it. If a real female comes by, he completely ignores her. She simply can't compete with the bottle's spell. In fact, there is practically nothing that will make male jewel beetles give up their beloved bottles. You need to physically remove them to stop the desperate mating attempt. Without human intervention the male beetles typically end up dying from dehydration under the hot sun or get eaten alive by ants because they so adamantly refuse to abandon their beloved.[1 2]

⋆

It's not so surprising that animals can be deceived by sexual superstimuli. This is actually just what we would expect. Because out in the wild there are two fundamental goals: to eat and to reproduce. And you only need food so that you have energy to stick around for sex and procreation.

Biological creatures have evolved this way because those who manage to procreate make the next generation. Let's do a quick thought experiment to illustrate: say we make two versions of an animal. The first version isn't interested in sex. Instead, it spends its time looking for food, sleeping and exploring its surroundings. When this animal eventually dies, it has not had sex and thus has left no offspring. So in the following generation no one has inherited its low sex drive.

The second version of our animal, on the other hand, is obsessed with sex. This animal also needs to obtain food, sleep and get to know its surroundings. It can't spend *all* its time and energy looking for opportunities to mate. But if it invests any spare time and energy into *mission reproduction*, it will of course leave more offspring than the sexless animal. Thus, given that sex drive, like most other characteristics, is partly heritable, the next generation will contain more sex-obsessed animals. And so evolution has played out since the beginning of multicellular life and sexual reproduction.

Millions of generations of selection for sexual desire has given us brains that are optimised to react to sex-related stimuli. Just like we are cued into food-related stimuli because that helped our ancestors survive. For a jewel beetle this means that shiny brown surfaces are appealing; for a male baboon it's a big red butt; and for humans it could be anything from a nice body or attractive face to more abstract qualities like thoughtfulness, confidence or power.

In any case, the first step in this inborn quest for sex is to find someone to have it with. This is a slightly more complicated – or advanced – deal for humans than it is for jewel beetles. The beetles pretty much just have to identify each other and then the deal is on. But humans have courtship, sometimes extensive, where we forge a connection and evaluate each other. As the 8 billion people on Earth attest, this has not historically held us back. But in fact, there are signs we are getting worse at pairing up. At least the share of adults who are single has never been higher and neither has the number of adults living alone.[3] This is the case in pretty much all developed nations but there is a twist. Take my native Denmark. The Danish numbers reveal that not *all* age groups are increasingly single. In the oldest generations the number of singles is lower than previously because people die later today and thus there are fewer widows and widowers.

Among people below the age of 50, though, singledom is breaking records each year. Of course, there could be multiple reasons for that. For one, we are richer today than ever before, so it is easier to live and get by on your own if you want. There is also less cultural emphasis on marriage, traditional values and the nuclear family. But when something impacts younger generations much more than older ones, it is tempting to suspect technology could play a role as well.

For instance, imagine you're a hopeful young man in medieval Britain. You've grown up on a farm in a small village and you need to find yourself a woman so you can get married and eventually inherit the family farm.

When you look around the village there aren't exactly limitless possibilities. In fact, there are precisely three women of appropriate age who are available. The first one is cute but doesn't like you. The second one is your cousin. And the third one is missing half her teeth.

Okay, that might be a little bit of an exaggeration. In reality, there would have been more options and you could always try looking at surrounding villages as well. Nevertheless, there's no doubt that your options would be rather limited. If you were a woman, instead, you might not even have had the option to pick. And for both sexes, your parents would have played a role in the choice too. In Britain, parents could reject a potential suitor if they believed that they didn't live up to expectations. But in other parts of the world arranged marriages were entirely normal (and in some places still are). So, if the village girl without teeth was the daughter of your father's business partner, you might have wanted to start saving up for a set of wooden dentures.

Compare this to a young person in the present day. Today, you might meet a partner at a party or in the bookstore holding your favourite book, *Super Stimulated*, in their hands. But the most common way that people meet today is through online dating. Not all countries have great data on this matter but we can see the rise of online dating in questionnaires from the United States. In 1999, 5 per cent of new American couples reported they had met each other online. In 2009, it was 22 per cent. In 2019, it was 40 per cent,[4] and today it is over half.[5]

The rise of online dating sped up in the 2010s with the popularisation of the smartphone and dating apps like Tinder, Bumble and Hinge. For our eager medieval farm boy, a few minutes on one of these apps would be an out-of-this-world experience. At the fingertips are thousands of potential partners, displayed in quick succession one after the other. Of course, not all of them would be interesting to him. And not all of them would reciprocate any interest he might have. But compared to the farm boy's situation in the Middle Ages, these apps would be a goldmine of options. 'A gift from above,' he'd probably call it. 'With so many options everyone must be able to find the right person! And people must be very satisfied with their relationships.'

Or?

If this was true, we would expect there to be fewer single people today, not more. But paradoxically, the plethora of options could be one of the reasons that fewer people are coupling up in the modern world. At least studies show that a variety of options tend to make us less decisive and more prone to regret our decisions, what is dubbed 'the paradox of choice'.

In one famous study, a group of psychologists conducted an experiment on this phenomenon using jam. On the first day of the study, they arranged a display of 24 different kinds of jam samples at a supermarket. Then on the next day, they repeated the display but only put six choices up. The psychologists discovered that the large selection of jam was better at capturing people's attention and making them try the samples. But it turned out that the smaller selection was more effective at getting people to actually make a choice and buy something. A variety of options created more interest, but fewer options were better at making people commit.[6]

You might recognise this from when you're trying to choose a movie to watch on Netflix and the endless options makes it difficult to choose, so you end up restlessly browsing instead.

American scientists have shown the same phenomenon is found in the dating world. In a study inspired by the jam study from above, scientists recruited unpartnered female students. They were told the university was developing a dating service which the women were asked to rate. Some of the women were then shown six potential partners, while the others were shown 24. And at the end, all the women were asked to choose whom among their options they liked the most.

The following week, the study participants were asked to return, and the scientists asked them to rate how satisfied they were with their choices. It turned out that the women who had been shown the most options also tended to regret their decisions the most. They were both less satisfied with the choice they made and more likely to change them when given the chance.[7]

**Too many fish in the sea?**

Even in the real world, there are signs that an abundance of dating options breeds indecisiveness and more unstable relationships. Just think about Hollywood marriages. Movie stars are the epitome of a desirable partner for many people: they are attractive, socially competent, rich and powerful. That means you probably won't find a group of people in the world with more dating options. Hollywood A-listers can pick and choose like no others and, in turn, their marriages are notoriously rocky.

Scientists have also noted this phenomenon among regular people. In a study on Danish marriages, it was found that people have a higher risk of divorce if they work in an industry with a higher proportion of the opposite sex. This was true for both genders but especially for men.[8]

★

Of course, even without a significant other, you can still have sex. For most animals, though, sex at least requires there to *be* another individual present. But as usual humans are creative.

An early sign of that dates all the way back to Stone Age Europe. Archaeologists around the continent keep finding so-called Venus figurines in their excavation sites. These are small figures that depict naked women, often with curves, large breasts and broad hips. For a long time, the purpose of these figurines has been debated: were they fertility goddesses? Or perhaps flattering attempts at self-portraiture? For obvious reasons, we cannot ask. But a likely explanation is that the figurines are simply the first example we have of pornography. Maybe something to bring on long hunting expeditions when your wife stayed back home. This hypothesis is backed up by the fact that the Venus figurines were clearly traded. The most famous figurine, the Willendorf Venus, was found in an Austrian village, approximately 100 kilometres west of Vienna. But it is not made from local materials. The limestone of the Venus of Willendorf is from a region hundreds of kilometres to the south. So most likely, the owner of the figurine was not the person who actually made it but someone who acquired it through trading.[9] Perhaps some person found out that porn can be a very lucrative business.

If so, they wouldn't be alone. In fact, the business case for pornography is so universal that it transcends species. In an American study, scientists trained our distant relatives, rhesus macaques, to illustrate this point. They exploited the fact that rhesus macaques *love* orange juice. So when scientists give them some liquid sunshine it is pretty hard to get it back. However, the scientists taught the monkeys that they could use their orange juice to trade if the scientists had other objects they would rather have. Mostly, the macaques weren't willing to give up their beloved juice. But it turned out the male monkeys *were* willing to trade for pictures of female monkeys' behinds.[10]

Back in our own species and time, we can see the draw of pornography in the fact that porn sites are some of the most visited websites in the world. After the Venus of Willendorf there has been a continuous escalation. First came sexual depictions in paintings, which gave way to the invention of photography. Photos escalated from pin-up girls to *Playboy* magazine, followed by porn videos, first as VHS tapes before today's explosion of internet-based porn streaming.

This is the well-known pattern we observed for food where, even thousands of years ago, we were already modifying the stimuli – for our food in the form of breeding sweeter fruits and fattier livestock. But modern technology has absolutely turbocharged the process. Modern internet-based porn makes use of a broad range of techniques to keep viewers watching. First, the models themselves are specifically chosen on their looks and sexual appeal. They are not average people but tend to have vastly exaggerated sexual characteristics. The women have large breasts and butts while the men have large penises. Female porn stars also tend to have low waist-to-hip ratios, which is a measurement associated with fertility. Studies show that men in general prefer a waist-to-hip ratio of 0.7[11] which is what is seen among porn stars but which is lower than what is found in the general population. Furthermore, the looks of the performers are also meticulously 'enhanced' using makeup and plastic surgery (which we will return to later). The whole thing is presented with all the tricks known from movies and TV production (which we will *also* discuss later). But briefly, lighting is carefully controlled, editing is used to remove any physical 'flaws' and to stitch together only the most appealing scenes. On top of that, changing camera angles and scenes introduce variety, which matters for satiation also when it comes to sex. In fact, variety is so crucial to keeping viewers watching, that an even more powerful trick is employed.

*

One day, former US President Calvin Coolidge visited an experimental farm with his wife.

The pair were given separate tours and at one point Mrs Coolidge paused at an enclosure of hens. She noticed the rooster was running around mating with one hen after another. Curious, she asked the farmer, 'How often does he do that?'

'Oh, dozens of times a day, Mrs Coolidge,' the farmer responded.

To which the First Lady remarked, 'Please tell that to the President.'

Later, when the farmer told President Coolidge the story, the president simply smiled and asked, 'Same hen every time?'

And the farmer promptly answered, 'Oh no, Mr President, a different hen each time.'

Then the President laughed. 'Tell that to Mrs Coolidge.'

This story is most likely an urban legend but it does align with a real psychological phenomenon observed in mammals. In a classic experiment, scientists put a male rat in a cage full of female rats in heat. The male rat doesn't waste any time and immediately begins mating with the female rats, often multiple times. In the end, though, he grows tired and retires to the corner of the cage to rest. Afterwards he refuses to repeat the act, even if the female rats approach him suggestively.

That is until scientists introduce a *new* female to the cage. Then the male suddenly regains his energy and hurries to mate with her. After a while, he grows tired anew and seeks rest but each time scientists introduce a new female, he springs into action. In fact, scientists can make the poor rat continue until he is on the verge of collapse. All it takes is a continuous supply of new females.[12] [13]

This phenomenon – that a new partner sparks renewed sexual interest – is named after the urban legend about President Coolidge and dubbed the 'Coolidge Effect'. In essence, it is the sex version of the buffet effect, which we've covered previously. Remember? In a buffet it takes us longer to feel full because the presence of variety delays satiation. Similarly, the presence of sexual variety delays sexual satiation for the male rat in these experiments.

As recent history suggests, the Coolidge Effect might apply especially well to United States presidents. But this effect is not *only* a phenomenon observed in the Oval Office and in rats. It appears in all mammals from sheep to hamsters and rhesus macaques. No one has performed an exact human replica of the rat experiment above, at least not with any scientific aim in mind. But several research groups have tried to simulate the concept in the laboratory. For example, researchers have shown male research participants porn films and used interviews or physical tests to register how sexually aroused they are during various scenes. In these studies, it turns out that men – just like male rats, rams[14] and macaques – are increasingly sexually aroused when the porn films introduce new females.

People have also studied whether the Coolidge Effect exists among female animals and women, but this is trickier. To put it bluntly, the line between sex and rape is not as clear-cut among many other mammals as it is among humans. A male rat, for example, doesn't wait for the consent of the female before it tries to mate. So, it can be difficult for scientists to gauge the preferences of the females themselves. Some scientists have tried getting at the question using special cages that connect via tunnels that fit female rats but not males because they are larger. In this way, the females can choose which male rat to visit and avoid being followed or harassed. Studies using this method suggest that female rats also show an interest in variety.[15] At least they will switch up their cages now and then. But the effect is much less pronounced than in males. This might explain why modern porn is especially appealing to men. Or at least why it is constructed the way that it is. You see, porn sites earn money by showing ads. That means they don't just have to attract viewers by using attractive models. They also need viewers to stick around as long as possible. The longer viewers are watching, the more ads you can show them and the more money you ultimately earn. That makes the Coolidge Effect valuable for porn sites. By strategically exposing heterosexual male viewers to many different females, the sites can postpone sexual satiation.

**Aren't we monogamous?**

The Coolidge Effect doesn't mean that humans are evolved to have multiple partners. It's just a good example that evolutionary adaptations are not organised or planned ahead. We also have adaptations that draw us in the opposite direction such as falling deeply in love and experiencing romantic jealousy. As a result, anthropologists have described various relationship structures in traditional peoples around the world, some of them alien to a Westerner. For instance, the Mosuo of China practise a form of fluid non-monogamous marriage where both men and women stay with their maternal kin all their lives. Sexual relationships occur through visits where a man goes to a woman's house and leaves in the morning. The relationships entail no obligations and can be had with many partners at once.

Another atypical example is that cultures in parts of the Himalayas used to practise a form of marriage where all the brothers of a family would share a single wife. This most likely evolved to make sure landholdings didn't have to be split between multiple male heirs.

The vast majority of cultures practise monogamy as it is known in the UK, though. The second most common relationship structure is polygyny where a single man has multiple wives. However, in these societies, polygyny is often restricted to the elite and most people still pair up monogamously.[16]

★

Over the past few decades, an increasing number of patients have sought help from doctors and psychologists with what they describe as 'pornography addiction'. Most of these patients are men, but there are also women among them. A few celebrity examples include the comedian Chris Rock,[17] the actor Terry Crews[18] and the singer Billie Eilish (who says that porn 'destroyed her brain').[19]

In 2018, the World Health Organization decided there was sufficient evidence to add 'compulsive sexual behaviour disorder' to what is called the International Classification of Diseases (sort of the official list of illnesses).[20] The disorder covers multiple forms of hypersexuality – an exaggerated focus on and desire for sex or pornography. In daily speech, you could call it sex or porn addiction. In professional settings, there is still debate on whether this is an addiction or rather what's called compulsive behaviour.[21] But for regular people, this distinction doesn't really matter. The point is that there are people who would like to stop watching porn or cut down on their viewing time but who are unable to. These people typically describe a need that can't be suppressed and desensitisation that makes them seek out more extreme content over time. Exactly like we have seen for food superstimuli earlier, where simple vanilla ice cream becomes boring and ends up as Ben & Jerry's with caramel sauce, chocolate chips, marshmallows, cookie dough and so on.

Of course, not everyone who watches porn becomes addicted. Just like we have established that we don't all react equally strongly to food superstimuli. Some lose all control and end up with a long list of health problems while others stay slim without a second thought. Most, however, are probably somewhere in between and this is likely true for sexual superstimuli as well.

A Danish survey of young people and their porn habits illustrates that point. In the survey, 59 per cent reported an occasional 'irresistible' urge to watch pornography. Half said that the porn they watch has become more extreme over time, and 43 per cent said they feel some degree of dependence.[22]

If we look back to Australia and compare ourselves to the jewel beetle, one could hypothesise that the draw of sexual superstimuli could lead to less sex in the real world. After all, the male beetles chose their beloved beer bottles over real females. Could the same thing happen to us?

To find the answer, we can look at surveys that attempt to map out how much sex people are having. In the UK, the US and Germany, these show the same result: we have less sex today than we used to.[23] [24] [25] People tend to begin having sex later – among Americans, there were two and a half times more virgins between the ages of 20–24 than there were among the same age group in the 1960s.[26] But even sexually active people have less sex and it is true whether they are single or in a couple.[27] It is difficult to say with certainty whether this pattern of decreasing sexual activity has anything to do with pornography. I would like to give you a precise answer, but unfortunately there aren't good enough studies in this area to say for certain. The few studies we have suggest that men in relationships who watch a lot of porn desire their partners less.[28] That could be a case of reverse causation, i.e. that men who don't desire their partners much decide to watch pornography instead. But it seems the porn habit usually precedes the decrease in sex drive.[29]

For now, we can't do much but speculate. As we've seen, humans are adaptation machines, and we are shaped by our experiences over the course of our lives. Today, many people will have most of their sexual experiences behind a screen. A young heterosexual British man can quickly see more naked women on the internet than his ancestors could see in the real world during their entire lives – combined. Even the mightiest emperor of the past with a full harem can't compete with 16-year-old Harry from Scunthorpe and his high-speed internet connection.

It is difficult to imagine that these experiences *don't* impact our sexual expectations and habits. As we discussed, porn stars are first highly selected for their sexual appeal and further technologically enhanced with the same aim in mind. Normal partners in the real world are forced to compete with these superstimulating opponents just like real natural food is forced to compete with scientifically optimised food superstimuli. It's easy to imagine that the sexual version of this battle ends as the food version so often does. But even in the absence of replacement, there is no doubt that pornography shapes the modern person's sexuality.

Sexual stimuli are highly rewarding and, when given the choice, animals will work hard to attain them. Early sexual experiences in particular seem to leave lasting effects and for this reason most research on the topic prefers to use young animals that have no prior sexual encounters.

For instance, American scientists have illustrated the power of early sexual experiences in a study on marmosets, small squirrel-like monkeys. The researchers applied a lemon scent to female monkeys and then let sexually naïve males mate with them a few times. Later, the monkeys were presented with the lemon scent in the absence of the females but had developed such a sexual attraction to it that they would get erections when they smelled it.[30]

The same thing has been demonstrated multiple ways in rats. Not only using smells but also using other neutral stimuli. In one example, scientists put female rats in little jackets and then let sexually naïve males mate with them repeatedly. Later, these male rats were shown to be more sexually attracted to females wearing the jacket than those without.[31]

You might recognise this type of conditioning as something similar to the one described by Russian scientist Ivan Pavlov in his famous experiment on dogs. Pavlov would ring a bell as he was feeding his dogs and showed that, eventually, the dogs would learn to associate the bell with the food. They would start salivating upon hearing the bell, even if no food was present. In other words, the innate response that dogs have for food – salivating – had been tied to something that was previously neutral to them. Dog trainers today use this effect by pairing treats with a clicker – a little device that makes a distinct clicking sound when pressed. Over time, the dog is conditioned so the click itself is rewarding, meaning the trainer doesn't end up overfeeding the dog but can make it feel good using something that was previously neutral.

Of course, this effect also exists in humans. For instance, it's why people in the advertising industry have always used the term 'sex sells'. Sexual cues such as scantily clad attractive people are inherently rewarding for us. By pairing them with a product, marketers hope to condition us with the product as the neutral clue that we build a preference for because we see it paired with a rewarding cue again and again. In a similar way it is unavoidable that pornography conditions the sexual arousal of viewers. Though, how it plays out is not known. We could speculate that neutral cues such as the screen itself would be linked to sexual arousal. But in any case, this effect would mean that porn is not just playing to human desires, over time it will also be forming them. The same is true for food superstimuli. We discussed how cuisines around the world all devolve into food superstimuli as countries get rich enough. But there remain differences in the taste preferences among cultures. You can think of these being driven by conditioning early in life where some of the tastes are the neutral cues that are paired with salt, sugar, fat and so on, eventually becoming rewarding on their own. Essentially, you are conditioned to like certain tastes that your culture exposes you to in early life. This conditioning is totally normal and doesn't only happen from superstimuli. Normal food and sex will do it too. It's just that superstimuli are especially potent.

# 10

# Shortcut to euphoria

So far, we've seen how multiple industries use the same approach to make money from superstimuli. First, identify something that humans are inherently attracted to. Then, make artificially exaggerated versions – superstimuli – that far exceed what you can find in the natural world. And finally, put on a monocle and laugh maniacally.

Okay, maybe not the last part, but there's no doubt that superstimuli are an extremely effective way to make money. It works for food (food superstimuli), and it works for sex (porn). But in fact, it is possible to simplify the approach even further.

You see, the superstimuli we have discussed so far are limited by the fact that they require the use of our inborn senses. For example, when you eat a piece of candy, the added sugar binds and activates taste receptors for sweetness on your tongue. This leads to nerve signals being sent to your brain, activating the reward system and making you feel a pleasant sensation. As we have discussed, there will also be signals sent to the brain from nerves in the gut and, while you're eating, your eyes, nose and even ears play a part too. In all these cases, the superstimuli have to be sensed and then the nerve signals have to travel to the brain. However, it is also possible to *directly* activate the brain's reward system without having to go through the senses. Typically this is done with a little help from the botanical world.

A good example is cannabis. Most of the physiological effects ascribed to cannabis are due to a molecule with the reader-friendly name Δ9-tetrahydrocannabinol.[1] It is often shortened to THC. When THC is absorbed by the body some of it ends up in the brain, where it can activate parts of the reward system and produce feelings of well-being or even euphoria.

More specifically, THC binds to structures on the surface of brain cells called cannabinoid receptors. You can think of this interaction like the one we talked about previously for artificial sweeteners. Imagine THC as a key that fits in a lock (the cannabinoid receptors). This fit activates the receptors and initiates various activities in the brain cells on which they sit.

Of course, we don't have cannabinoid receptors in the brain for them to be activated by cannabis. The fact that THC fits in these receptors is coincidental. The receptors are actually there to react to certain molecules that your brain produces itself. These compounds are called endocannabinoids and are usually released in connection with rewards – when you experience something enjoyable. They're involved in mood, appetite and memory. However, THC activates the cannabinoid receptors much more potently than your natural endocannabinoids do and THC also sticks around in the brain for longer. This is why it causes such intense effects. It's like a chemical shortcut to superstimulate yourself.

As in our previous examples with nicotine and cocaine, the intensity of the effect from THC partly depends on how quickly the molecule reaches the brain. If you eat cannabis, THC has to pass through your digestion first and will then be metabolised in your liver. That means it takes anywhere from

30 minutes to a few hours before it reaches the brain in high amounts. But if you smoke cannabis instead, THC reaches the bloodstream almost immediately and hits the brain in a few seconds.[2] This is why smoking is the most popular way to use cannabis: there is a shorter time between stimulus and reward. But besides this, the strength of the THC effect also depends simply on how much you consume. The more THC, the greater the effect. So, can you guess how cannabis has evolved over the past few decades?

In my native Denmark, police routinely screen cannabis confiscated on the streets for its THC content. Here are the average values for different years:[3]

1970: 0.6 per cent
1980: 1–2 per cent
2000: 8 per cent
2017: 25 per cent

Do you recognise the trend? This is the exact same pattern we have discussed previously for ice cream eaters and pornography viewers: desensitisation. Over time, we lose sensitivity to the stimuli we are routinely exposed to, meaning we need to escalate to stronger versions if we want to feel the same reward. And just like with ice cream, pornography and all the other superstimuli, there is no chance that the escalations end here. Today, cannabis cultivators have already bred cannabis plants that contain 34 per cent THC.[4] [5] But even that pales in comparison to the THC products that people have started using in certain parts of the world.

To understand what has happened, we need to go back to 2014. On 1 January of that year, the American state of

Colorado legalised cannabis for the general public. Since then, multiple states have followed suit so that cannabis is now legal in roughly half of US states. The movement has also spread to Canada, and other countries such as Mexico, South Africa, Thailand and Germany are taking steps in the same direction. Legalisation doesn't just change the law, though, it also changes the kind of people involved in an industry. Before legalisation, the selling of cannabis was mostly the domain of petty criminals. But upon becoming legal, a new crop of operators arrives. These include experienced business people, young ambitious graduates looking to build their career and various investor types with briefcases and complicated spreadsheets. As it turns out, these kinds of people are a lot better at doing business than random street criminals. It naturally limits consumption of a product when the people charged with selling it have a hard time getting out of bed in the morning. In contrast, the new players in the industry have wasted no time applying strategies that are all too familiar to us at this point. One is that they have been hard at work in an effort to stuff as much THC into their products as humanly possible. After all, THC is what people are paying for and what gets them coming back. So if you want to gain customers and beat the competition, you'd better escalate the superstimuli. Cannabis companies do so by making various oils, waxes and crystals that can be anywhere from 50 to 95 per cent pure concentrated THC, often consumed using sleek, colourful vaping devices.

The result is that cannabis users today take larger doses of THC than previously.[6] But unfortunately that doesn't just intensify the cannabis *experience*; it also intensifies the *side effects*, because most of the side effects of cannabis also stem from

THC. Case in point: cannabis is known to decrease sperm quality, and the higher the THC intake, the worse sperm quality gets.[7] [8] The same is true for mental side effects such as memory problems and for the risk of getting a psychosis.[9] [10] [11] [12] [13] The latter happens in genetically vulnerable people, where THC ingestion can induce delusions, hallucinations and paranoia. If one is lucky, symptoms pass but they can also turn into full-blown schizophrenia, which causes difficulty differentiating between what is real and what isn't.[14]

Finally, there is also a paradoxical increase in the risk of developing anxiety and depression from using large quantities of THC.[15] Many people might use cannabis to try to avoid these very ailments. But again, this is an example of the body being difficult to trick. If you push it one way, it tends to push back just as hard. In the short term, cannabis reduces depression and anxiety by activating the reward system. But in the long run the brain adapts. So imagine that you have anxiety. On a scale from one to ten, you might usually be a seven – quite anxious – so you start smoking cannabis to feel calmer. Smoking brings you down to a three on the anxiety scale. But as you keep smoking your brain adapts and, among other things, reduces the activity in the part of the brain that is activated by cannabis. This means that cannabis might now only bring you down to a six. And in the absence of cannabis you'll be even more anxious than you were before. Maybe an eight or nine. That means, you'll have to increase the THC dose, and if you take the drug out of the equation, you'll feel even worse than you did to begin with.

★

In addition to steadily increasing the amount of THC in their products, cannabis companies are also working to make them more attractive in other ways. Just like food companies, they use additives such as artificial flavours and aromatics to enhance the experience. They research new solvents and techniques to make the THC absorption quicker. And they are competing on all the usual business premises: delivery, customer service, price, accessibility and so on.

This kind of competition might benefit you when it's your local clothing store or electronics shop striving to offer you the best experience. But when it comes to superstimuli, it's not a good thing that providers are well-oiled machines. We know that our consumption of various 'vice products' is affected by how easy they are to obtain.

The best parallel to cannabis is the Prohibition era in the United States from 1920 to 1933. During this time, alcohol was made illegal following campaigns by religious groups and women who hoped to decrease alcohol-related domestic violence. If you've heard about Prohibition, it is probably as a case study of a failed policy. It made people brew homemade spirits instead, some of which could blind the drinker. And it also created a black market for alcohol, which gangsters like Al Capone used to make tons of money. In the end, Prohibition was rolled back democratically even though it had been implemented that way in the first place.

Something rarely mentioned about Prohibition, though, is that it actually worked. We can't say exactly how much alcohol people were drinking while it was illegal. But there's no doubt that it was less. We know that because rates for alcohol-related illnesses fell sharply. For instance, fewer people were getting liver cirrhosis, fewer people were diagnosed with alcohol-related psychoses and even arrests for drunkenness and disorderly conduct decreased. When alcohol was legalised again, it is estimated that Americans only drank half as much alcohol as before Prohibition. And alcohol consumption only returned to previous levels decades later, in the 1960s and 1970s.[16] [17] [18]

The reverse pattern can be seen for sports gambling, which was illegal in the United States until the Supreme Court struck down the ban on a federal level in 2018. Since then, 38 states have legalised it and studies show both the number of gamblers and the amount gambled has since increased in these states compared to the non-legal states.[19] At the same time, so has the number of personal bankruptcies and rates of domestic violence.[20]

Of course, both these examples don't have to imply that banning something is the right thing to do. They simply illustrate that there is a trade-off between accessibility and consumption. The fewer barriers and frictions there are between people and a 'vice', the higher the consumption. This is beneficial to remember on a personal level. If you, for example, want to lose weight, the best thing you can do is to decrease the accessibility of food superstimuli. Remove them from your home and put barriers and friction between you and them.

For society as a whole, the trade-off becomes a philosophical question on balancing freedom and health. If we return to cannabis, proponents have tried to argue that no trade-off exists – the argument for legalisation has often been that people will find ways to get their drugs, whether they are legal or not. Trying to deter them would then just be a waste of resources and playing straight into the hands of criminals.

However, we can now see what actually happened in Canada and US states that allow recreational cannabis.[21] In the US as a whole, cannabis use has dramatically increased since legalisation efforts began and is now at an all-time high.[22] However, the increase has been higher in states where cannabis has been legalised.[23] To get a sense of the impact, we can compare two similar states such as Colorado – the first state to legalise cannabis – and Minnesota. The two states have similar populations, geographic sizes and economies. Back in 2014, cannabis use in the two states was also close to equal, but since legalisation, Colorado has outrun Minnesota, and residents here now consume 24 per cent more cannabis.[24] The same is true if Colorado is compared to other similar states such as Kansas. Critics of these comparisons have often noted that they rely on questionnaires. You can probably imagine that even if you ask people anonymously, some might be hesitant to admit doing something illegal. Still, all other evidence points in the same direction. For example, in states where cannabis is legal, we see more cases of cannabis use disorder, more cases of cannabis poisoning among young children and pets, more arrests for cannabis-impaired driving (and relatedly a rise in the number of traffic accidents).[25] [26] [27] [28] [29] [30] [31] And finally, when people in emergency rooms do urine tests, more of them are positive for THC than before.[32]

So, we can't get around the fact that legalisation increases consumption. This also makes intuitive sense. If you want to drink less alcohol, you'll do better living in Saudi Arabia than in the global West. And if you want to eat less junk food, you'll be more successful in Ethiopia compared to the United States.

**Big Tobacco's playbook**

In many ways Big Tobacco created the playbook that cannabis companies are now using. Tobacco companies cultivated plants with greater levels of nicotine to hook their users, in the same way that cannabis companies increased the THC content. The trick of using additives to improve the taste of the smoke also came from Big Tobacco, as did the idea of adding chemicals to expand the airways and to help the brain more rapidly absorb the addictive substance.

However, the connection between the tobacco industry and the cannabis industry doesn't stop there. Internal documents show tobacco companies have been eyeing up cannabis as a business opportunity since the 1970s. By now, multiple tobacco companies have made acquisitions and investments in cannabis companies, hoping to benefit from this type of smoking as well.[33] This isn't a new strategy but an approach Big Tobacco has repeatedly used since governments and consumers started cracking down on them. Back in the 1980s when the government crackdown intensified, various tobacco companies began buying up food manufacturers. Kraft Foods – think Cadbury, Toblerone and all kinds of cookies – was acquired by the largest tobacco company, Philip Morris. And Nabisco, which makes Oreos, among other things, was also bought by the same company and later sold to the second largest tobacco producer, RJ Reynolds. Turns out the expertise gained from making addictive products in one area is transferable to others.

# 11

# The happiness plant

Five thousand years ago, a group of people lived in what is now Iraq who called themselves sag-giga – 'the black-headed people'. Their neighbours called them Sumerians, though, and that's the name we use today. The Sumerians were the first real civilisation on Earth. They built cities with thousands of inhabitants that had large pyramid-shaped temples where priest-kings reigned. These cities were larger than anything the world had seen before, and so, to make things run smoothly, the Sumerians also blessed humanity with the world's first bureaucracy.

We shouldn't fault them too much for this. The Sumerians also gifted us the world's first written language, cuneiform, to make up for it. Here, symbols were carved into wet clay tablets which were left to dry in the sun. This made the tablets into permanent documents which are so durable that thousands of them have survived to this day. For instance, if you've read *Jellyfish Age Backwards*, you might remember the *Epic of Gilgamesh*, the oldest known written story in which a king leaves his people on a quest for immortality. Or maybe you have heard about the infamous Sumerian merchant Ea-nasir whose shady business practices made him the subject of the world's first customer complaint. The thousands of tablets mean that linguists have managed to decode the Sumerian language, so we have a unique insight into the world's first civilisation. We can read personal letters, prayers to the gods, business agreements, laws and recipes for food and medicine. The medicinal tablets are particularly interesting for you and I because here we find a plant that the Sumerians called hul gil. A literal translation would be something like 'happiness plant'. In English, we haven't been quite so straightforward in our naming and simply call this plant the opium poppy.

The opium poppy is a 30–80-centimetre-tall plant with large, beautiful flowers. However, the look of the flowers is not the reason humans grow opium poppies. We don't grow them for food either, though poppy seeds are used in bread and pastries. No, the major reason we grow opium poppies is in the name itself. They contain *opium*, a euphoria-inducing substance that comes from the liquid inside the seed capsule. If you tap the liquid, you can process it into a sticky black mass that induces an intense state of pleasure, as well as a feeling of comfort and lack of pain.[1 2]

Opium works this way because of a mechanism resembling the one we discussed for cannabis. The plant contains certain compounds that can activate receptors in brain cells of the reward system. These are not *cannabinoid* receptors, but *opioid* receptors. Again, they aren't called this because the brain is made to react to opium. Opioid receptors exist because they are naturally activated by certain molecules that are released by the brain in connection with reward. These include, among others, enkephalins and endorphins.

The natural pairing of the brain's own opioid chemicals with opioid receptors helps us to regulate pain and our moods. So, when opium activates these chemicals artificially, this becomes yet another way to superstimulate ourselves.

*

We don't know exactly *why* the opium poppy contains opium. It is not vital for the survival of the plant, and the substance is only produced for a short time. Some believe that opium protects the poppy from insects and predators.[3] But this would be strange because wild poppies *don't* produce opium. Suspiciously, opium only exists in cultivar species.

So another possibility is that the opium poppy is a clever parasite. Or, an organism that exploits us for its own benefit. By producing a euphoria-inducing substance, the plant ensures that humans will take care of it and help to spread it. Historically, pretty much all opium farmers were addicted to their crops. And not just economically because the plant was their source of income. Rather, opium farmers used to be physically dependent and got that way from inhaling opium during the harvest or absorbing it through their skin. Some of them were likely also users of the finished product because opium dependence used to be the second most frequent form of addiction, after alcoholism. It was described by the ancient Greeks and Romans but was never as serious in Europe as in the Middle East and Asia where opium was even *more* prevalent than alcoholism. Some of the opium users ended up with their addiction due to thrill seeking just like today. But people could also become addicted because opium was by far the best-known pain reliever for much of history. So back in time, chronic pain could pose a tragic dilemma: escape your hellish pain but risk ending up in addiction hell instead. Doctors, alchemists and others spent thousands of years trying to work around this problem by finding a painkilling but non-addictive form of opium. As so often in this book, the first breakthrough came from a German chemist in the nineteenth century. Friedrich Sertürner succeeded in isolating the most important active compound in opium. He called this new compound *morphium*, after the Greek god of dreams, Morpheus. Today, we know it as *morphine*.

Sertürner hypothesised that morphine would be less addictive than opium because it is concentrated and thus five to ten times stronger. This means that you can use less substance for a similar effect, and it was Sertürner's hope that this would equal less addiction.

To test his theory, Sertürner began to sample his own drug. As you might have guessed, this didn't end particularly well. Today we know that increased strength *increases* addiction risk, and Sertürner found out the hard way. So, he went from celebrating his new discovery to sending the world a warning, writing, 'I consider it my duty to attract attention to the terrible effects of this new substance I called morphium, in order that calamity may be averted.'[4]

Unfortunately, though, the world didn't listen. Over the following decades, tons of opium users escalated to morphine, sinking them even deeper into addiction and despair.

However, before long help was on the horizon. This time from a Scotsman called Alexander Wood. Wood helped to invent hypodermic needles, which are the needles we use today for injections. They revolutionised medicine in that drugs could now be administered directly into the bloodstream. As always, though, inventions aren't all good or bad. Alexander Wood believed that his needles could also be used to prevent morphine addiction. He argued that you could *only* become addicted to a drug if you took it orally. By *injecting* a drug, addiction would be impossible.[5]

So once more, Wood's theory was a good example of Niels Bohr's famous quip: it's difficult to make predictions, especially about the future. In fact, this theory was so profoundly wrong that it surely tops some kind of list. You see, when you inject a drug directly into your bloodstream, it reaches the brain more quickly compared to if you swallow or smoke it. So instead of curing the problem, the new injections made morphine even more addictive and just increased the problem.

Despite the misfire, though, hope wasn't vanquished. At the end of the nineteenth century, scientists at the German pharmaceutical company Bayer conducted experiments with morphine and vinegar acid and discovered that the combination produced a new related substance. After a brief study they concluded that this new substance had a different chemical structure from morphine, and therefore *couldn't* be addictive. In fact, they believed that the new drug would be a *cure* for morphine addiction and that it could be used to treat everything from coughs to pain.

The excitement around the new drug was so great that the pharmaceutical company decided to give it a glorious name. They narrowed the potential names down to two options: a name inspired by the German word for miracle, *Wunder*, and a name inspired by the German word for heroic, *heroisch*. The company chose the latter and dubbed their new drug . . .[6]

Heroin.

Again, as you might have guessed, things didn't go exactly as planned this time around either. Heroin reaches the brain *even* faster than morphine and is twice as potent. So in some way, I guess you could say that Bayer were technically right in claiming they could cure morphine addiction. Because as soon as those with a morphine addiction tried heroin, they switched to the stronger stuff right away and never looked back. Thus, it wasn't long before the addiction problem spiralled out of control and Bayer voluntarily recalled the drug. Later, heroin was declared illegal, and the monumental failure was enough to deflate the research balloon.

For a time.

Some scientists kept looking, but at this point it had become established in the collective consciousness that drugs like heroin, morphine and opium are addictive and require caution.

That is until the 1990s when the American pharmaceutical company Purdue Pharma launched a new pain relief medication called OxyContin. The active ingredient in OxyContin is called oxycodone and is an opium-based substance just like morphine and heroin. Oxycodone itself was nothing new. In fact, it had been discovered all the way back in 1916.

No, the new thing here was – say it with me – that Purdue Pharma had discovered a way to make it *impossible* to get addicted to the substance.

The company's trademarked Contin-system, as they named it, ensured that the drug was absorbed slowly and gradually. That should be sufficient to prevent addiction, right? Wrong. In no time, patients started chewing the pills to get a full dose at once. Others dissolved the pills in water and injected them into their bloodstream.[7] But there were also patients who took the pills exactly as directed but still became addicted.

OxyContin was a gold mine for Purdue Pharma, and soon enough, other pharmaceutical companies hopped on the bandwagon. They made their own opium-based pain relief medications, and things turned out the same as previously. The opium-based drugs started an epidemic, just like earlier waves did. This epidemic is called the Opium Crisis and has hit especially hard in the United States where it has cost over half a million lives.[8]

No, that is not a typo.

Unlike the previous examples, though, this part of the opium story isn't about ignorance or wishful thinking. Because Purdue Pharma knew from the beginning that their products could and were being abused. We know this because they have been sued so many times that we can read their internal emails and documents, including correspondence in which the company's own scientists warn that OxyContin can be addictive and has been demonstrated to be so in the laboratory.

The company, however, chose to leave that out of their marketing.

*

The story of Purdue Pharma is a good illustration of the types of people that superstimuli industries often end up attracting.

The company is owned by the American Sackler family, who entered the pain management sector after one of the family members got wind of a common misunderstanding among doctors. Richard Sackler was had studied medicine himself and noticed that many doctors incorrectly assumed that oxycodone was weaker than morphine. The opposite is true – oxycodone is stronger – but Sackler also discovered an explanation for the myth. Many American doctors associated morphine with life-threatening operations and dying patients and were therefore cautious about prescribing it but many didn't know much about oxycodone.

'It is important that we be careful not to change the perception of physicians,' wrote one of the Purdue employees in an internal email.[9] Because if the misunderstanding continued, doctors could be convinced to prescribe Purdue Pharma pills to treat all kinds of ailments, including ones that doctors would never prescribe morphine for.

Like all pharmaceuticals, OxyContin needed to be approved by the authorities before it went to market and this is often a lengthy and thorough affair. The doctor who oversaw the approval process, Curtis Wright, happened to get a well-paid job at Purdue Pharma just one year after the approval. (Wright has since denied any impropriety in this process.)[10]

With the approval in place, Purdue Pharma was ready to market their drug. They hired an army of sales representatives and rewarded high sales figures with cars and exotic vacations. To aid sales, the company hosted hundreds of seminars on pain management for doctors, of course with all expenses paid. As the leadership noted in internal emails, this approach was particularly effective when the seminars were held over the weekend. According to the company's own internal data, that made doctors subsequently began to prescribe twice as much OxyContin as before.[11] Purdue Pharma even began sharing free samples of their medicine with patients.[12] That's the same business model we know from lowly drug dealers where you try to get the customer addicted, so they'll come back for more. The first 'free' round quickly pays for itself.

Of course, the rollout of OxyContin wasn't without criticism. It didn't take long for doctors in the most affected parts of the United States to notice the havoc it caused in local communities. There were reports of waves of addiction and criminality, the likes of which had never been seen before. Before long some of the addicted people also started dying. Opioids like oxycodone, morphine and heroin aren't just dangerous because of the horrors of being addicted. They are also dangerous because opioids have many functions in our bodies, including regulation of our breathing. So, if you overdose on opioids, you stop breathing.

Unfortunately, the doctors who raised concerns about OxyContin weren't met by a receptive public. Purdue Pharma had foreseen the critique and, like others in this book, taken cues from the tobacco industry. The company sponsored leading scientists to sow doubt about the concerns raised by doctors. The most prominent of these, Russell Portenoy, called the criticism of OxyContin hysterical and dubbed it 'opiophobia'. Another prominent scientist, David Haddox, was more inventive. He admitted that OxyContin made patients demonstrate signs of addiction, but – they weren't actually addicted! Rather, these were symptoms of what Haddox termed *pseudo-addiction*. A syndrome he came up with that could easily be mistaken for addiction. The key difference, as he explained, was that pseudo-addiction was caused by taking *too few* opioids. In other words, these patients didn't need to get off OxyContin. They just needed higher doses, and then they would feel better.

As the opioid crisis spread, it eventually became impossible for the champions of opioids to keep maintaining these claims. Many have since gone back on their words. Portenoy, for example, later admitted some of his previous statements were false and not based on data. He later testified against the company, though that is obviously too late for many of the victims.

The opioid crisis that started with OxyContin and other forms of pain relief medication has only escalated since. Mexican narcotics cartels were not going to stand idly by while others earned this much money in their area. So, they pulled a trump card from their deck: a drug called fentanyl. Fentanyl is an opioid, just like morphine, heroin and oxycodone, and it is used by doctors for management of severe pain and for anaesthesia. You can see it as the next level on our escalation ladder because fentanyl is 25–50 times stronger than heroin. The cartels realised they could use this to their advantage by luring in people already addicted to OxyContin or other opioids with an even stronger fix. That is extremely risky for the drug taker, though. Fentanyl is so strong that 2 milligrams is enough to kill a person. That means a kilo of fentanyl is enough to take the lives of 500,000 people. But even this potency is not the end of the escalation ladder. Some addicts have already progressed to another synthetic opioid called carfentanil, which is 100 times stronger than fentanyl and which was originally developed to sedate elephants.

It might be difficult to understand how those with substance use disorders can't see this is a direct path to the abyss. But opioid addiction is impossibly difficult to shake. First, there are the extreme withdrawal symptoms. Those who try to wean themselves off the drugs begin to sweat profusely, have muscle aches and stomach cramps and will experience uncontrollable vomiting and violent diarrhoea. In fact, withdrawal symptoms can even be fatal. So, people not only return to their drugs because of an addiction; they will do whatever they can to avoid the withdrawal symptoms.

Unfortunately, we build a tolerance to opioids incredibly fast, which means that people need to take higher doses or stronger drugs to avoid the withdrawal symptoms. But then they risk overdosing, and as previously mentioned this has already cost the lives of over half a million Americans so far.

★

The story of Purdue Pharma doesn't have a happy ending. The company reached a settlement requiring it to pay a fine worth billions, but the Sackler family are still billionaires and by paying the fine they secured legal immunity from future civil lawsuits as part of the settlement.

Still, this story is helpful. Because combined with what we've learned about other industries that sell addictive products, it can help us to understand how companies that profit off compulsion and addiction keep criticism at bay.

The playbook is always the same. To avoid being held responsible you start by adamantly denying that your product is in any way addictive. Even when hundreds of thousands of people were already addicted to OxyContin, Purdue Pharma maintained that addiction was rare. Just as the tobacco industry used to argue that smoking wasn't addictive, because, as they would say, cigarettes aren't intoxicating like alcohol or drugs – so they simply can't be addictive, right? People can easily smoke a cigarette and then continue with their day. Moreover, smokers who quit don't experience life-threatening withdrawal symptoms like alcoholics or heroin addicts do. So, there is no way that smoking could be addictive.

We are still in this phase, the denial phase, when it comes to many of the products mentioned in this book. As we've covered, we are now debating whether people who watch porn for hours a day are 'addicted' or 'compulsive'. And when it comes to food superstimuli, we are still debating whether something as fundamental as food can be addictive.

In this phase, the debate usually ends in semantics. *What do we mean by 'addiction'?* These kinds of word games can derail the debate for a while, but they rarely last. As we learn more about the underlying biology, it becomes increasingly impossible to deny the problem. For instance, as scientists learned more about nicotine's effect on the brain, the tobacco industry was forced to capitulate and admit their products are addictive.

At this point, we enter phase two, the admission phase. Here, the company confesses – yes, some people develop a substance use disorder. But the problem isn't the company and its products. The problem is the people who develop a disorder. They are addicts, who are guilty of their own ruin. Richard Sackler from Purdue Pharma used to say that addicts *want* to be addicts. In internal communication, Purdue Pharma listed 'key messages that work' to promote their opioids, including 'it's not addiction, it's abuse' and 'it's a matter of personal responsibility'.[13] [14]

One of the older members of the Sackler family, Arthur Sackler, had a similar defence when he was accused in the 1950s of making people addicted to another drug – a sedative called Valium (the apple doesn't fall far from the tree). Arthur Sackler claimed that it wasn't the drug or his dubious marketing that was wrong. The people who developed a substance use disorder had to have made the mistake of mixing the drug with alcohol or cocaine. We see the same strategy for food companies such as Coca-Cola, which used to sponsor an organisation called the Global Energy Balance Network. Officially, this organisation's goal was to research obesity, but the studies always concluded that the cause of obesity was a lack of exercise.[15] It wasn't Coca-Cola's fault that people were gaining weight. Some people are just lazy, and that was why they gained weight. Today, we know the opposite is true – diet rather than exercise is to blame for our weight issues.

When the companies switch to the strategy of shaming consumers, they run into a problem, though. Because these consumers typically make up the largest part of the company's earnings. The 10 per cent of the population that drinks the most alcohol buys more than half of all alcohol sold.[16] So, if everyone who drinks too much suddenly stopped, the alcohol industry would immediately go under.

In the same way, 10 per cent of the people who gamble at casinos make up nearly 80 per cent of casinos' income.[17] Among mobile gaming apps, 10 per cent of customers make up nearly two thirds of the companies' earnings.

This exact problem is exemplified by Purdue Pharma themselves. When public pressure grew too great, the company agreed to transition to a version of OxyContin that was more difficult to abuse. It didn't work that well, but the small added hindrance made sales immediately fall by over 25 per cent when people with a substance use disorder switched away. They made up such a large part of the business that revenue was significantly hit by chasing them away.

# 12

# Dopamine and what you need to know about it

In the 1950s, Canadian scientists Peter Milner and James Olds were looking for the brain's reward system. They knew such a structure had to exist somewhere in there. The question was *where?*

To find the answer, the scientists designed an experiment that has since inspired thousands of dystopian science fiction stories. Their idea was to use electricity to localise the reward system by activating different parts of lab rats' brains. Because brain activity is simply that: electric signals in the brain. We have no idea how you get from electrical signals between brain cells to the feeling of something being pleasant. But Milner and Olds reasoned that if they could activate the right brain cells electrically, they could make the animal feel good and reveal to them that this was the spot.

You can picture the experimental setup as an even more simplified way to produce a superstimulus than taking euphoria-inducing drugs. If you take heroin, for example, it needs to travel to your brain after being injected into the bloodstream. Afterwards it binds to receptors on brain cells in the brain's reward system, and only then do the electrical signals start. But by inserting an electrode directly into the brain, you can bypass that small delay by activating the brain cells of the reward system pretty much instantaneously.

The Canadian researchers designed their experiment as follows: rats had electrodes implanted in various parts of their brain and at various depths. Then the rats were placed in cages where they could activate their own electrode by pressing a small pedal. Rats are inherently curious so when you put them in a new cage, they immediately begin to investigate their environment, sniffing and feeling it. In Olds and Milner's experiment, this meant that the rats would eventually press the pedal and activate their electrodes. In most cases, not much more happened after that. A few of the rats even started avoiding the pedal, meaning it must have been an unpleasant experience to activate whatever part of the brain their electrode had been inserted in.

For four of the rats, however, it was a different story. After these rats experienced the electric stimulation, they pretty much ceased all activities besides frantically pressing the little pedal again and again to make themselves feel good. The scientists had hit the reward system.[1]

Since the 1950s, similar experiments have been repeated countless times, and it isn't unusual for the lab rats to get so caught up in their electric self-stimulation that they press the pedal more than once a second. They become so addicted that they are even willing to expose themselves to pain to gain access to the self-stimulation. For instance, by running across an electrified grid that shocks their feet. It is also typical for the rats to lose interest in everything else around them. Even if they are starving, the rats ignore food that is put right in front of them and choose to self-stimulate instead. So if scientists don't intervene, the rats will starve to death even if food is readily available.[2]

The reason electrical self-stimulation is so addictive is because it carries virtually no delay. We have learned that the more rapidly a drug reaches the brain, the more addictive it is. This is why people often inject drugs directly into their bloodstream, so that they aren't delayed by the digestive system. But not even needles can compete with electrical self-stimulation. Here, the only delay after pressing the button is from the current running through the wire and into the brain's reward system.

Neuroscientists have illustrated this principle by conducting versions of the rat self-stimulation experiment in which they create an artificial delay between pressing the pedal and the electrical signal. When you do that, the lab animals become less addicted to the self-stimulation, exactly as expected.[3]

*

Soon after the Canadian rat experiments, American scientists in New Orleans started conducting similar experiments on people.[4] The scientists, who were led by Robert Heath, were working with patients in a psychiatric hospital. Back in the 1950s, there weren't the same rules or oversight for scientific research as there are today, so the scientists had close to free rein.

In the beginning, the New Orleans scientists were only using brain-implanted electrodes to measure brain activity in their patients. But over time, they also began to electrically stimulate different parts of the patients' brains, just like in the rat experiment. The hope was that this electrical stimulation could be used to cure illnesses like depression and schizophrenia which, it was reasoned, involved maladaptive brain activity.

Robert Heath and his colleagues claimed that the electric stimulation could get patients to feel everything from fear to joy, depending on which part of the brain was stimulated. They went especially far with one patient. We don't know the name of this man who was simply called 'patient B-19'. But he was 24 years old, white and had experienced a turbulent childhood with a violent father. B-19 didn't have any close friends growing up, had switched schools nine times and finally dropped out of high school. After that he developed an addiction to drugs including amphetamines and one day he was arrested for cannabis possession. Since then, he had lived at the psychiatric treatment facility, where he was also diagnosed with clinical depression and epilepsy.

B-19 also happened to be homosexual. And for some reason *that* was what the doctors decided they would try to treat. With B-19's consent, they implanted electrodes into multiple parts of his brain. The idea was to activate them as he watched films of heterosexual sex to train him to associate this with pleasure. The doctors also created a device so that B-19, under observation, could activate his own electrodes by pressing a few buttons. One of the electrodes was in the brain's reward system and soon enough B-19 was pushing it feverishly. Robert Heath describes how the patient would press this button over a thousand times in a three-hour session and how he protested and pleaded with the scientists when they took the device away.

We don't know the fate of B-19 but he was not the last person on whom electrical self-stimulation was used. It has been trialled particularly often on patients with chronic pain. Here, the theory has been that electrical self-stimulation could disrupt the feeling of pain. Because if brain activity is electrical signals, then maybe stimulating the right place would offer the patient relief from their pain. A few doctors in New York tried this as they were treating a 48-year-old woman who suffered from chronic back pain. The doctors had tried everything from pain medication to surgery, but nothing helped. In a last ditch effort, they implanted electrodes in the brain of the woman and, as with B-19, she was given a device to activate them whenever her back hurt. However, after the operation, the woman reported that one of the electrodes made her feel sexually aroused whenever she pressed it. As with B-19 this electrode was located in the brain's reward system. But as opposed to B-19, the woman didn't only have access to the electrode while under observation. The idea was that she should be able to activate her electrodes whenever she felt pain, so she was allowed to take it home with her. As a result, it didn't take long for the woman to develop a crippling addiction to self-stimulating. She abandoned everything else in her life so that she could press the button all day long. Besides the button-pressing, she became completely inactive and neglected both her family and her personal hygiene. Over time she developed an open sore on her finger from pushing the button so much, but even this couldn't make her stop. She even attempted to take the device apart to increase the intensity of the stimulation.[5]

Based on this description, you might think that this woman was existing in a kind of pleasure nirvana. But that's not how she described it herself at all. The woman described how the stimulation gave her a strange underlying sense of anxiety, and she began to experience frequent anxiety attacks and exhibit signs of depersonalisation. In the laboratory, the doctors could see that the level of stress hormones in the woman's blood jumped as soon as she began to self-stimulate. And she was begging her family members to remove the device and hide it from her. But each time they did so, she was soon struck by an irresistible urge to find it again.

To understand this reaction, it's helpful to know about the American psychologist Kent Berridge and his work. Berridge is a neuroscientist who has thoroughly described how the feelings of *desire* and *pleasure* are not in fact the same, even if we often think of them that way. In fact, the two are distinct from a neurological perspective. Berridge says that *wanting* isn't the same as *liking*.[6] Of course, the two things are often closely related. You want an ice cream, and it feels nice to eat it. But it turns out that they can also be entirely separated from each other.

For example, people with a heroin misuse disorder experience great pleasure the first few times they take the drug. But as we know, the brain adapts and as the person continues their drug use, the reward decreases. Eventually, heroin might not even result in a good feeling at all. But that doesn't mean the desire disappears as well. On the contrary, it only gets stronger as the addiction grows. So, the person with the disorder ends up in a situation in which they experience a desperate *wanting* for their drug but no pleasure or *liking*.

Both humans and animals in the aforementioned electrode experiments end up in the same situation because the circuit that is stimulated in their brains has to do with *wanting*, not *liking*. From the outside, it might seem like these patients – and lab rats – are enjoying the stimulation. They keep pushing the button, after all. But what they experience isn't pleasure at all. It's just an overwhelming desire to push the button. That's why the self-stimulating woman became stressed out and anxious, not blissful, by pressing the electrode button. She also described a feeling of wanting to have sex, but the stimulation never led to pleasure or orgasms.

This distinction between wanting and liking can help us clear up a few things about the superstimuli we have discussed so far. Take food superstimuli. We have established that fast food, supermarket snacks and so on are designed to be hard to resist. You might think this means that I'm arguing a McDonald's burger is the height of culinary delight. But that is obviously not the case. The number one meal for *liking* in your life was probably some fabulous meal at a family-owned Italian restaurant or a home-cooked meal your grandma used to make. No, fast food and other superstimuli are not optimised for *liking*. They are optimised for *wanting*. That is, they are made to create an urge in you. The same is true for sexual and drug superstimuli. It is not as if pornography is more pleasurable than actual sex or that sitting alone on a couch smoking cannabis is something most people would actually prefer to do compared to living their life out in the world.

Most of us are simply used to thinking about desire and pleasure as one and the same. I'm sure you've heard ignorant comments along the lines of: 'She keeps going back to him, even though he abuses her. She must like being hit,' or 'He keeps drinking even though it is tearing his family apart. He must value alcohol more than the well-being of his children.'

These statements would be true if desire and pleasure were always in tune. But they aren't. This discrepancy causes problems in many fields of science such as philosophy and economics in which influential theories presume that desire and pleasure are one and the same.

Take, for example, the economic theory of 'revealed preference'. This theory is meant to explain why our actions don't always match our words. One person might claim to care about animal welfare, but they still buy eggs from hens in cages. According to the theory, this person 'reveals' that their words are empty, because otherwise they would be willing to pay a little extra and put their money where their mouth is.

The theory does work in some cases, but it falls apart in many of the areas we deal with in this book.

For instance, do overweight people 'reveal' through their choice of food, that they value unhealthy food over their health?

This doesn't align with the fact that half of all adults try to lose weight every year. It also doesn't make sense, if you think about how much money people are willing to pay for weight loss courses and diet products. In fact, it is possible that overweight people get *less* pleasure from their food than others because of desensitisation (and the shame that comes with eating what they want to).

*

*Wanting* is associated with a molecule in the brain you might have heard of before: dopamine. Only a small portion of your brain cells release dopamine, a few hundred thousand out of 86 billion. Nevertheless, this molecule is enormously important for your behaviour.

Dopamine-producing brain cells exist in a few small regions in the middle of the brain, right behind your mouth. For us, the most important area has the lovely name the ventral tegmental area. From here, dopamine-producing brain cells send small projections to other parts of the brain. One key target is an area called the nucleus accumbens, which is a central hub for processing reward and motivation. Other targets include the frontal lobes, which are responsible for planning, problem solving and self-control and the hippocampus, which, among other things, plays a role in memory. Together, we call these circuits the brain's reward system.

You have probably heard dopamine described the reward substance. Something akin to 'Social media is cheap dopamine,' or 'Chocolate gives me a huge dopamine hit.'

But dopamine isn't actually what gives you a feeling of pleasure. We have other substances to thank for that: endocannabinoids and opioids, which we've already looked at. The brain cells that release and respond to endocannabinoid and opioid brain cells sit in the same region as the dopamine-producing brain cells and are closely related to them. So we can still talk about the brain's reward system. The point is just that dopamine isn't the reward itself.

We can see this in experiments where researchers reduce the amount of dopamine in an animal's brain. Using a chemical method, scientists can kill the dopamine-producing brain cells, so that the amount of dopamine drops to 1 per cent of its natural level. This makes mice lose the desire to eat. They basically stop moving even to seek out food. Without intervention, they would starve – even if food is readily available.[7] Importantly, however, that is not because the mice have lost the ability to feel pleasure from food. If scientists inject a sugar solution directly into the mouth of the mouse, its face will light up and it will begin to lick its nose – a sign that it likes the sugar.[8] The mouse has just lost the desire and ability to seek out the pleasure.

This passiveness is typical for dopamine-depleted animals. They don't make spontaneous movements and they lose their curiosity and their desire to explore their environment. If you put them into a new cage, they will just sit there looking at you.

We also see this phenomenon in people who have sustained damage to the parts of their brains where dopamine-producing cells are located. For example, if a blood clot kills brain cells in this area. Such patients can develop a condition called abulia, which means 'lack of will'. They are typically inactive, lying in bed all day, responding to questions with monosyllabic words like 'yes' and 'no'. They also lose interest in the things that they used to care about. But if you give them dopamine-increasing drugs, they liven up and regain some of their old personality.

Another example is Parkinson's disease, in which some of the dopamine-producing cells in the brain begin to die. These cells are not a part of the reward system, but located in a nearby brain structure called the substantia nigra. This means 'black substance' because the area is darker than the rest of the brain due to a high concentration of a pigment called neuromelanin, which is a byproduct of dopamine production.

The substantia nigra is essential for our ability to activate and control our movements. So when these brain cells begin to die, patients get tremors, slow movements, muscle stiffness and balance issues. To reduce the symptoms, Parkinson's patients are typically given medication that increases the levels of dopamine in the brain. This can help the remaining dopamine-producing cells make up for the ones that have died. However, the increase in dopamine is not limited to the substantia nigra. The medicine raises dopamine levels all over the brain, including in the reward system. If doses are too high or patients are genetically vulnerable, this can trigger addictive or compulsive behaviour.

Some can develop an uncontrollable sex drive, beginning to watch large amounts of pornography, obsessively seeking out sex with their partners or sex workers or experience changes in sexual desires.[9] Others start overeating, develop an exaggerated need to shop or suddenly develop gambling addictions.[10]

*

All our organs are complex but the jobs of many of them are relatively easily understood. At a glance, the role of the heart is to pump blood around the body and the job of the kidneys is to filter waste products and excess water from the blood. In comparison, the task of the brain is being a biological supercomputer and it does such a good job that it has somehow attained consciousness and begun studying itself.

For that reason, it isn't so shocking that neuroscience is the most complex field of medicine. While engineering students might shed a tear at their thermodynamics exams, aspiring medical and biomedical students get teary eyes when half of their exam questions are long Latin words for obscure brain areas. So, if you think this stuff about the brain is difficult to grasp or you need to read it twice, you are not alone.

Well, that was a pause for a deep breath. Now let's see if we can understand the dopamine system a little better. The first thing you need to know is that dopamine-producing neurons release dopamine in two ways.

The first can be described as a constant low-level secretion. Here, dopamine-producing brain cells constantly shoot a little dopamine to nearby brain cells. This dopamine release sets the bar for how easy it is to perform various actions: the more dopamine, the fewer obstacles. Imagine various parts of your brain are constantly making suggestions on what you should do at any given moment. One region might say: 'We should eat some chocolate.' Another might say: 'We need a nap.' And a third might say: 'That book deadline is coming up soon . . . maybe we ought to sit down and write.' You obviously can't do everything at once, so a special area in your brain is responsible for choosing between all these different options. This area is called the basal ganglia and if we were to use an analogy, you could picture it as a kind of doorman who decides which potential action is allowed access to the muscles so they can be carried out in real life.

The suggested actions don't all appeal equally to the doorman but vary in intensity depending on the state of the body. If you've just eaten, the chocolate voice will be a faint whisper that isn't heard. But if you're ravenous, it'll be shouting and screaming. This increases the likelihood that the doorman will let the suggestion pass to the muscles, so you start eating.

But in addition to the intensity of the suggestion, there is also a general threshold for action, as mentioned. This threshold is set by dopamine. The more dopamine there is in the 'doorman' part of the brain, the lower the bar will be to let suggestions pass.

This mechanism can be most easily understood if we see what happens when you increase the amount of the dopamine in the system. We've already learned that mice with a dopamine deficit become extremely passive. This is because the low level of dopamine sets the bar for action unreasonably high. Using our doorman analogy, low dopamine levels make the doorman set sky-high demands. So even if parts of the mouse's brain are shouting, 'We need to eat!' or 'Move, god damnit!' they won't be allowed to pass and the mouse ends up doing nothing.

The opposite occurs for people who take cocaine or amphetamine. Both drugs temporarily increase the level of dopamine in the brain, *lowering* the bar for actions to happen. Using our analogy, the doorman becomes much less selective and starts letting all kinds of suggestions pass. That's why cocaine and amphetamine make people lose their inhibitions. They talk more than usual, say whatever comes to mind, become hyperactive and behave impulsively – even violently, if that's the impulse that arises.

**Trapped in their own bodies**

In 1982, six people in California were found suffering with the same mysterious symptoms. They were all unable to move or speak but were still alive. At first glance, the patients had nothing to do with each other and doctors feared a new illness might be spreading. But later, it was discovered that all six patients shared the same drug dealer. He had sold them what he believed was the heroin-like MPPP. However, it turned out that the chemist producing the drug had made a terrible mistake. Instead of MPPP, he had accidentally made a similar compound called MPTP, which is toxic to dopamine-producing brain cells. So the six Californian patients weren't moving or speaking because all their dopamine-producing brain cells were dead. They had completely lost the ability to initiate actions.

The patients were flown to Lund University in Sweden, where they received an experimental treatment, but unfortunately it didn't work. To this day, we haven't discovered a way to help them, and the unhappy ending to this story is that they never recovered. They spent the rest of their lives trapped in their own bodies, unable to initiate actions.[11]

*

The second way that dopamine-producing brain cells release dopamine is through a process called phasic transmission. It has this name because the brain cells produce a bunch of dopamine at once that quickly disappears again. You can think of it like a dopamine peak, after which levels normalise. This form of dopamine release doesn't have anything to do with picking between different actions. It has to do with learning *new* ones. The peak is a signal that what you just did was good, and that you should repeat it in the future if you face a similar situation.[12]

We can understand this part of the dopamine system by seeing it in action when it's most important: in early life. Imagine yourself as a baby, sitting in your highchair with a bowl of purée and a baby spoon in your hand. You're not very good at controlling your arms yet, so you flail wildly and most of the purée ends up all over the place. But at some point, you are lucky and manage to move your arms just the right way so that you hit the target: your mouth. Immediately you taste the yummy purée, a nerve signal travels from your tongue to your brain and there is a dopamine peak in the reward system.

Again, dopamine itself *isn't* the reward. The reward comes from other molecules and dopamine doesn't tell you whether something is good or bad. Rather, dopamine peaks deal with expectations. They arise when something is better or worse *than expected*. And 'baby food in your mouth' feels nicer than 'no baby food in your mouth'.

The dopamine peak ensures that the brain activity that took place right before the reward has a greater likelihood of being repeated in the future. That means that as you become better at getting food into your mouth, the connections in your brain that help you put the spoon in your mouth are strengthened. Eventually, you can do it without a second thought. When this happens, the dopamine peaks disappear. Baby food still tastes good to you, but it is no longer an unexpected reward. You have learned that the reward will come when you activate the connections in your brain that make you first put your spoon into the purée and then move it to your mouth.

However, one day, the baby food gets switched out with Coco Pops. You've never tasted them before, but the cereal is more rewarding than the purée. So when the Coco Pops hit your tongue, you get an unexpectedly large reward and dopamine is released. Now you learn to build on what you already know. The new lesson is that it is better to put Coco Pops in your mouth than baby purée. So, if you encounter Coco Pops in the future, there is an increased likelihood that you'll put them in your mouth to get another reward.

Later, your parents might trick you and put Brussels sprouts on your plate. They taste bitter and because bitterness is often associated with various toxins, we tend not to like bitter flavours, particularly as children. So when you dutifully eat your first Brussels sprout, there is *less* of a reward than expected. In this case, there's no dopamine peak. Instead, the dopamine level in the reward system *falls* below normal. This makes it less likely that you'll put the Brussels sprouts in your mouth the next time you see them.

To sum up, dopamine fluctuations in the brain are part of learning. When something feels better than expected, they train the brain to repeat the action so that you can get the same feeling again. This system is enormously clever in the wild, where it helps us zero in on the most calorie-rich food and the best mates to help us pass on our genes.

But in the modern world these mechanisms go off-course and make us seek out 'empty' rewards like junk food, pornography and cigarettes because they have been sneakily designed to activate the brain's reward system to an unnatural high degree.

*

As dopamine teaches you to associate certain actions with a reward, something interesting happens, which backs up that dopamine is about learning. The dopamine peak starts to happen sooner and sooner when you face a potential reward. Take our Coco Pops example. In your first encounter, the dopamine peak will happen when you taste the Coco Pops. But later, it will begin earlier. Initially, it might move back to the point where you put your spoon into the cereal. Then later, it will move back to when you see your parents pour the cereal into your bowl or when they open the pantry where the box is sitting. The reason this happens is that your brain moves backwards from the reward when it's learning. Initially, you learned that Coco Pops give you a good feeling when you put them in your mouth.

But as soon as that's taken care of, your brain begins to learn what step precedes the spoon manoeuvre. Or what makes the Coco Pops appear in the first place. So, you get a dopamine peak if you ask your parents for Coco Pops and they give you some or if you someday find the box in the pantry and pour yourself a bowl. In both cases the reward still doesn't happen before the Coco Pops touch your tongue. But your dopamine peaks happen as soon as you're sure that a reward is on its way. In this way your brain builds up the action pathways that are required to get rewards by working backwards from the reward itself. This also means, however, that the introductory step is part of creating a desire to continue the action sequence. You might start desiring the Coco Pops as soon as you see the box, or even when you're just passing the pantry.

Perhaps you can see how these mechanisms are one of the reasons desire and pleasure are not the same. Because desire arises at the first sign of a possible reward. It can even pop up if you know there is something unpleasant standing in between you and the reward. Say, drawing harsh smoke into your lungs or putting a needle into your arm. For rats, desire can even be activated by touching a rod that gives them an electric shock, so long as the reward system is electrically stimulated immediately afterwards. Then the rats learn that the shock precedes the reward, and they begin to desire the shock even though it is unpleasant, because the reward follows.

*

Dopamine and its role in learning is why people with substance use disorders are not only plagued by the substance itself but also various associations that they've subconsciously developed in connection with their addiction. The mere sight of a needle or certain environment can trigger immense *wanting* because their brains have learned that these cues are part of the early steps in the action sequence that result in a reward.

This is why people who want to stop taking cocaine usually need to stop partying entirely. They might manage to control their urges in everyday life, but the brain has learned that 'alcohol' and 'loud music' precede cocaine. If they go to a party like in the old days, this triggers urges that are nearly impossible to resist. You might recognise this from social smokers too. They don't have problems avoiding smoking in their daily life. But as soon as they begin drinking, the urge appears, and before long they've got a cigarette in their hand.

Again, this is because their brains have learned to associate alcohol with the reward from smoking, and so the urge appears whenever they've been drinking.

We've observed an extreme version of this phenomenon in soldiers returning from war. When the American army conducted surveys during the Vietnam War, it turned out that 43 per cent of soldiers had taken heroin during their deployment. Half of those who had tried the drug admitted to having an addiction.

But when the soldiers returned home there was no national heroin crisis, as one might have feared. In a follow-up survey, only 1 per cent of the soldiers remained addicted.[13] The same was observed during World War II, in which both the Allies and the Axis Powers gave their troops large quantities of amphetamine to make them more resilient, aggressive and attentive. Between April and June 1940, for example, the Germans sent out 35 million methamphetamine pills to their soldiers. But again, there was no great addiction crisis after the war.

This might be hard to believe. We've seen how things transpired in other situations in which many people have tried an addictive substance. In the case of OxyContin – which is weaker than heroin – there was a health crisis involving millions of people. How could the soldiers have avoided addiction?

One possibility is that the soldiers exclusively used the drugs to alleviate the psychological toll of being at war. After the war ended, they no longer needed these chemical crutches and could continue without them. We know, however, that many soldiers continue to have mental health disorders when they return home from a conflict zone. Moreover, it is not a requirement to feel bad to develop a substance use disorder.

This also happens to a great many people who are thriving and doing well.

The explanation more likely has to do with what we've learned about associations. After the war, the soldiers were transported thousands of kilometres back home to an entirely different environment. They changed their social networks and their daily lives all at once. Hence, the cues – 'triggers' if you will – the soldiers had learned to associate with their drug disappeared. They were like party smokers who never went to another party and so never had the urge to smoke again.

Of course, it's not particularly practical to ditch your friends, change careers and leave the country when you want to shake your Pringles habit. But you can make do with less. The story of the soldiers in the Vietnam War and World War II suggests that the first step in gaining control over your bad habits is removing as many triggers as possible. This is why I previously highlighted the importance of not having unhealthy food lying around if you have problems with overeating. Because the mere sight of the bag of candy or chips often acts as a trigger. It is easier to win the battle if you avoid it in the first place.

In addition to avoiding triggers, one could also imagine fighting addictions by 'retraining' the brain, so it learns that the triggers *don't* predict that a reward is coming. This is significantly more difficult, and we don't know much about how to do this. We know from people with alcohol use disorder that it works, though.

Alcohol affects the brain in multiple ways, including that it causes opioids to be released in the reward system. Therefore, people can use a pharmaceutical called naltrexone to block opioid receptors, which helps them 'delearn' the addiction. When naltrexone blocks opioid receptors, alcohol loses its ability to release rewards in the brain.[14] Users will take the medication in the morning and if they fall into temptation later and have a drink, it will be a boring experience. The reward will be less than expected. And this experience, as you might remember, would make dopamine levels fall, 'teaching' the brain that alcohol isn't a reward worth expending energy to seek.

Perhaps something similar could be used to manage other forms of addiction. If your Coco Pops suddenly begin to taste like Brussels sprouts, it won't take long for the desire to decrease. However, the problem is that the brain isn't so easy to trick. You might find some way to make your snacks less tasty. But when you know deep down why they aren't as appealing anymore, the method is unlikely to work. Significant research is being undertaken in this field, though, so one day we might be able to train ourselves out of our addictions.

**Molecule nicknames**

We have a habit of giving nicknames to many hormones and molecules in the body. For instance, in this book we've established that dopamine is more of a 'learning molecule' than a 'happiness molecule'. However, biology isn't always so easily sorted into boxes.

Dopamine has many roles besides the ones we've talked about. Outside the brain, it affects urine production in the kidneys, it can activate immune cells, and it also has a role in the development of our eyes.[15] [16] [17] Similarly, we've learned that opioids make us feel pleasure, but are also used to control our breathing. A third example we haven't yet encountered could be the 'love hormone', oxytocin. This hormone influences our feeling of attachment to other people. But it also makes the muscles around the uterus contract, which means that it can be injected to induce labour.

If we humans had designed our own bodies, we might have streamlined things and used a single molecule per function. But biological organisms are spaghetti machines, and everything is reused in the most bizarre ways. Keep this in mind when you read about complex biological phenomena: it's not only about the molecule but also about where, how much and in what context.

# Part 3

## SCREEN SUPERSTIMULI

# 13

## *Friends*, just in real life

If you think about it, humans are strange animals. Most other mammals have four paws, fur and a tail. But here we are walking around on two legs, tailless while only covered by hair in a few strategic places.

We are also the only animals with a chin; among the few animals that can sweat over their whole body; and we have, to put it mildly, atypical dietary habits that include drinking the milk of other mammals and using fire to cook our food.

Still, we can't get around the fact that the most remarkable and unique feature of humans is our huge brains and resulting intelligence. In fact, one could argue that the human brain is the most complex structure in the entire universe. We might be learning about some of its weak points in this book but, weaknesses aside, it is hard not to be amazed by human ingenuity. I'm currently writing this chapter 10 kilometres above the Atlantic Ocean in a giant metal tube with wings that burns million-year-old plankton to stay airborne. And my tool is a machine that uses sand and electricity to perform mathematical operations. In comparison, the grandest achievement of chimpanzees is licking on a stick and using it to retrieve insects.

Of course, no matter how intelligent any single human is, none of us can construct an aeroplane or a computer on our own. Humankind's great strength isn't just raw intelligence, but the combination of intelligence and strong social instincts. This duo allows us not only to have great ideas and learn about the world but to collaborate on making new discoveries and to build on each other's knowledge. In any case, though, our sociality is just another way humans are distinct from most other animals. Many animals live solitary lives and only meet up with their fellow species members to mate with them or to fight them. Even the animals that tolerate each other's company usually live in small flocks.

You and I are also members of small 'flocks' such as our families and friend groups. But we also participate in enormous flocks like our nation states and, increasingly, global society as a whole. This is only possible because we aren't just tolerant of each other's company. No, we have an instinctual need for companionship, validation and acknowledgement from other people. As you might have guessed by now, though, that means you can create superstimuli for these needs as well. In fact, one could argue that our social instincts are the most exploited of all.

*

One of the best places to witness our strong social instincts is in the fiction we like. Consider how many television series that can be boiled down to a plot like this: a family or friend group go about their lives and have experiences together.

Such series are not some epic tale of a journey to the ends of the world or a magical universe, but a depiction of a group of supposedly normal people, going through their daily lives and having fun. Examples include *Seinfeld*, *Friends*, *How I Met Your Mother*, *The Office*, *Modern Family*, *The Big Bang Theory*, *New Girl*, *The King of Queens*, and many, many more.

Obviously, although these series are supposed to be about normal life, they are not realistic. The characters are meant to represent regular people but tend to live in large houses and apartments, be attractive and funny, have practically endless free time and experience much more than normal people in the real world. At the same time, TV series make sure to cut out all the mundane aspects of daily living such as transportation and waiting time, nights where nothing happens, repetitive tasks and boring work. In essence, you could say a TV series is an exaggerated version of real life. Just like strawberry-flavoured candy is an exaggerated version of real strawberries.

You will even recognise many of the methods used to create these 'social' or 'experience' superstimuli, from all the other superstimuli we have gone over so far. At the core, the method remains doing experiments on the user – viewer in this case – to learn how to most efficiently hook them in. Historically, TV series have had to rely on manual methods such as using focus groups who watched the episodes and gave feedback. This is still done today, but in a digital world, there are also much more efficient methods available. For instance, Netflix is famous for extensive use of what is called A/B testing.[1] This is a method for optimising digital products and consists of splitting users into groups that are shown different variations of a feature, design or piece of content to learn which one works best. For instance, you could test what colour 'buy' button on your website leads to most sales by showing some users a green button and some a red button while tracking the conversions.

Of course, the A/B testing at companies like Netflix is not restricted to simple questions like button colours or to two different options. It is used on a massive scale and applied to virtually every aspect of the platform to optimise user engagement. It is the basis for the autoplay feature, for the 'skip intro' feature and for all the design choices of the site. Moreover, A/B testing is also used to personalise the site to keep each individual user watching. Netflix might test several versions of a thumbnail for the same show, each tailored to different audiences. Someone who has watched a lot of action movies might see an image of explosions, while someone whose watch history suggests an interest in dramas might see a tender moment between the main characters.

Admittedly, Netflix are not alone in this type of optimisation; it is the name of the game online. Practically all major websites you use are doing live experiments on you to improve their retention rates (including porn sites).

Some content producers even use experiments to guide the creation of the content itself. A good example is the British company Moonbug Entertainment, which is behind a string of wildly popular YouTube franchises for young children such as CoComelon. As reported by the *New York Times*, the shows are optimised for maximum engagement using the following setup: young children are placed in front of a screen showing an early version of a potential cartoon. Behind the screen is an even bigger screen which the company calls the Distractatron. This screen displays a continuous video loop of various everyday activities such as scenes from coffee shops and hair salons. The child is monitored while watching the cartoon and every time they look away from the cartoon show, it is recorded. Once this kind of experiment has been done with many different children, the company can identify the least engaging parts of their cartoons and change them until the children stay attentive throughout the whole thing.[2] If you've ever seen kids watch these shows, you'll know how well the experimentation works, creating an almost zombie-like devotion. The same thing could be said about adults watching Netflix for hours on end. Netflix are so successful they make up 15 per cent of *all* global internet traffic at this point.[3] They know they have the public hooked. In a now famous quote, the founder, Reed Hastings, said the following in an investor meeting when he was asked whether the company was concerned about the competition from other streaming services: 'When you watch a show from Netflix and you get addicted to it, you stay up late at night. We're competing with sleep, on the margin. And so, it's a very large pool of time.'[4]

*

Of course, the logical part of our brains knows that shows we are watching on a screen are not the real world. But that doesn't necessarily apply to our subconscious. You might, for example, have experienced a slight sadness when you finish a long television series, perhaps feeling almost like you've lost someone or something.

The reason we feel this way is because we develop something called parasocial relationships with characters we follow on the screen. This phenomenon was first described by American sociologists in the 1950s when people started having televisions in their homes.[5] The sociologists described how some viewers formed relationships with the fictional characters as if they were real people. A few even described falling in love. This happens because we cannot fully distinguish what is real and what isn't. To part of our brains, the people on the screen are our real friends and the experiences are our real experiences. So when we get to the end of a series, it can feel like 'losing' friendships.[6] Something similar is happening on a large scale today when people get invested in the lives of influencers and celebrities on social media and YouTube. Even though this relationship is also pretty much only one-way, the subconscious can interpret it as something real and form parasocial relationships.

The problem with all of this is that our parasocial relationships end up competing with the real thing. It's totally normal for people to sit behind a closed door and follow along with the lives of their parasocial friends and family while their *actual* friends and family members are just around the corner. This would seem quite puzzling to someone with no preconceived notions. Imagine that you and I are aliens from a distant galaxy that just discovered Earth. Our civilisation is curious about the new planet, so we are sent down to study its inhabitants. We receive special instruction to study Earth's dominant species, humans, and to describe their behaviour. What would you report when you're observing a person who sits pretty much motionless a whole Sunday looking at a screen?

Human scientists seeing a mouse do something similar would group this behaviour with the rodents doing electrical self-stimulation. Even though something is happening on the screen and in the brain of the viewer, from the outside all you see is an animal remaining completely passive for hours on end while receiving external stimulation.

We'll encounter this principle again and again in this part of the book. For the companies that earn money from manipulating our social instincts, it is all about keeping our attention and making us spend as much time as possible with their product. As a consequence of this, some start to neglect the real world.

*

When European explorers and traders first travelled to India, they noticed something strange. There were striking similarities between some of the Indian languages and the ancient Greek and Latin that was taught back in Europe. There were even similarities between the Indian languages and the explorers' mother tongues. For example, the sixteenth-century Italian merchant Filippo Sassetti remarked that words such as snake, god and various numbers were strikingly similar in Sanskrit and Italian.

When news of the linguistic commonalities reached Europe, curious linguists began studying them. They concluded that there was only one logical explanation: Indian and European languages had to have a common ancestor. A primitive language that was the origin of them all.

Today we group all these languages as 'Indo-European languages' because we now know these early linguists were right. Languages from English and Latin to Hindi, Urdu, Bengali and Sanskrit are descendants of the same ancient language. This language was spoken by a small group of tribes that spread their language by brutally conquering large parts of the known world.[7] The tribes left no written record, so we don't know what they called themselves. But today we call them Indo-Europeans or Yamnaya. Originally, they lived on the vast grass steppes of southern Ukraine and Russia about 5,000 years ago, where they led a nomadic existence. The Yamnaya people were the first people to tame the horse, and from horseback they herded cows, goats and sheep, which were their livelihood. This nomadism offered much more freedom than the settled farming practised by the rest of Europe at the time. But the Yamnaya way of life was also incredibly violent. The tribes of the steppe were brutal like few others.

We know this from one of their most important traditions – what was called *kóryos*. This was a coming-of-age ritual to mark the transition from childhood to adulthood. A little like Confirmation or Bar Mitzvah today. Except that the kóryos involved bloody sacrifices to the gods and painful rituals. The purpose was for the young men to temporarily take on a new identity based on the wolf. They were to become violent, impulsive, cunning and cruel. The new identity permitted the kóryos group to exist outside the normal morals of society. They were allowed to steal, plunder and kill – so long as they didn't go after members of their own tribe.

After the rituals, the young men were kicked out of the tribe's territories to fend for themselves for a while. They had to hunt wild animals to survive, and as part of their new wolf identity they were to attack and plunder other tribes – stealing livestock, killing local men, taking children as slaves and forcing women into marriage or worse.

If the kóryos men were successful on their expeditions, they could return to their tribe after a few months or years and be reintegrated as respectable citizens. But given that success also yielded livestock, wives and a band of brothers, they could also choose to form a new tribe. In both cases, a new generation of young warriors would soon see the light of day and then the story was bound to repeat itself.

Some of the kóryos groups likely perished in their attempts to attack other tribes. But we know that many were successful warriors because we can see this in the archaeological record,[8] where items related to the Yamnaya spread from their homeland on the steppe and in a few centuries reached the furthest corners of Europe to the west and the Indian subcontinent to the east. We can also see the conquests in the cultural footprints left by the Yamnaya in the societies that followed them. Their gods became the Greek, Roman and Norse gods and are the reasons why these mythologies have so much in common. And their ways continued in traditions such as those of the Vikings whose expeditions are reminiscent of the kóryos tradition. The Vikings used shock troops called berserkers, who were warriors dressed in wolf- or bearskins that rampaged against the enemy in an uncontrolled, trancelike and bloodthirsty haze. Across Europe, we also see traces of the old ways in stories and legends. Think about the myths of werewolves – men who in the light of the full moon transform into wolfish creatures that terrorise everything around them. Finally, we know of the success of the Yamnaya because we can see it in our DNA. If you are ethnically European, Central Asian, Iranian, Afghan, Pakistani or Indian, these warriors were some of your ancestors. In both Europe and the Indian subcontinent, the Yamnaya warriors encountered societies of farmers who weren't necessarily peaceful themselves. But these farmers were powerless in the face of the ruthless horseback warriors. In some places the farmers were completely wiped out while in others some of their genes remain to this day, likely from the people who were enslaved or the women who were forced into marriage.

★

Europe and the Indian subcontinent aren't the only places with violent pasts. In the rest of the world, we see other groups conquering each other in similar ways. At the same time as the Yamnaya people were spreading from the steppes of Ukraine and Russia, another group was doing the same in Africa. We call these tribes the Bantus, and they originated in present-day Cameroon but managed to spread across large swathes of the continent using similar methods as the ones employed by the Yamnaya.[9]

Fortunately, things are different today and you can take a trip to the supermarket without worrying about being ambushed by a group of wolfskin-clad men on horseback. But all over the world we still witness kóryos-like groups spontaneously emerge when conditions are right. You see, prehistoric people didn't invent a novel form of behaviour here. It is part of human nature that young men are the most aggressive and impulsive members of society. In fact, some psychologists use the term 'young male syndrome' to describe how young men are overrepresented in most statistics relating to violence or risk-taking.[10]

Around the world the social group with the greatest risk of being convicted of violent crimes is men in their late teens and early twenties. This group is also overrepresented among gang members, suicide bombers (the average suicide bomber is a

20-year-old man),[11] [12] in car accidents[13] and in people visiting the hospital after punching a wall.[14]

Don't get me wrong, this doesn't mean that all young men are violent maniacs. Most of us have managed to avoid becoming gang members and have never subjected an innocent wall to violence. But my point is that there are certain biological impulses among young men (probably because of testosterone) that increase the risk of this type of behaviour. These impulses were encouraged in the past through traditions like the kóryos, while today we do what we can to redirect them into more productive and peaceful activities. Historically, the most successful of these redirections has been sports, which George Orwell described as 'war minus the shooting'. Here, you can run and fight over a ball and everyone is allowed to return home alive and well.

With technological developments, there are also new ways to redirect violent impulses, though. Most interestingly, computer games.

*

There are computer games about everything between the heavens and the Earth. But in many of the most popular games, it's difficult not to be reminded of the kóryos traditions. You play on a team, and the goal is to kill the opponent or beat them in some kind of competition. For this reason, computer games have often been accused of contributing to violent behaviour. After all, it's not hard to imagine that people who get used to plundering and killing behind a screen would become more likely to do the same thing in real life.

Over the past few decades sociologists have been studying whether this connection is true. But although they have searched thoroughly, there is no evidence that violent computer games make people more violent in real life.[15] [16] We could reach the same conclusion by looking at society. As previously mentioned, violent games are incredibly popular and only becoming more so. If they made young people more violent than previously, it would be clear in crime statistics. But in fact, crime is *dropping*. Fewer young people are convicted of violent crimes today compared to ten and 20 years ago.[17]

Paradoxically, this *fall* in crime might be a result of the very games that were accused of inspiring criminal behaviour. We can see a clue to this in an interesting Dutch study. Here, scientists researched how criminality in a given country was affected when a new version of *Grand Theft Auto* was released there. In case you haven't heard of *Grand Theft Auto*, the premise is that you play a criminal that goes on various missions that include beating people, robbing them and committing murder. So, if there was ever a time when crime should rise due to violent video games, it should be when a whole country suddenly starts playing *Grand Theft Auto*. In the Dutch study, however, it turned out that crime rates *fell* when *Grand Theft Auto* was released.[18]

American scientists have observed the same phenomenon when violent movies are released. They used movie violence indices and crime statistics to study whether violent movies inspire violent crimes. But again, the answer was no, the violent movies made crime rates *fall*.[19]

Can you guess why?

The reason is really simple: you cannot be in two places at once.

If you are pummelling some poor guy in a computer game, you can't simultaneously be beating someone up in the real world. So as screen time rises, there is less time to do violent crime. There is less time for any real world activity, really. This is remarkably clear in surveys from my native Denmark. Every fourth year the Danish government commissions an extensive study of the health and well-being of young Danes. Included is a survey where one of the questions is whether the person answering has spent physical time with a friend outside of school, sports or work during the last week. In 2009, 63 per cent of Danish men aged 15–19 answered yes. Today, the number is 38 per cent.[20] This fall is evident for all age groups and both genders. And spending time with friends is not the only real life activity on the decline. So is going to parties and attending sports events and concerts. That is not to say young people don't socialise or have experiences anymore. It's just that socialising increasingly happens online and that experiences are also increasingly digital, for instance in the form of movies and computer games. After all, can your real friends *really* compete with the most entertaining parasocial 'friends' on YouTube, TikTok and Netflix? Are the real life activities available to you really as engaging as hyperoptimised TV series and computer games? Answers will differ depending on who you are, but all the numbers suggest that the digital world is winning this fight more times than not and that many people are increasingly living their lives in the digital world. In general, people spend more time alone and at home than ever before. In data from the United States, people report spending 40 per cent less time with other people than they did just ten years ago.[21]

This change might also explain some of the other social problems we have already discussed. Remember, for example, that more people are single than ever before, and that we also have less sex than people of the past, even in relationships. Previously, we discussed how dating apps and pornography probably contribute to this problem by causing choice overload and high expectations. But another possible explanation is that we simply spend less time in the real world and interact with fewer people. On a positive note, that means less violence. But it also means fewer positive interactions in the form of friendships, relationships and sex.

# 14

## Scrolling into infinity

In a large British study, scientists tried to map out what makes us happy in our daily lives. They designed a smartphone app that would ask thousands of study participants throughout the day what they were doing as well as how happy they were at that point in time. Unsurprisingly, people reported feeling especially happy after they had sex, after they had been social or after they had participated in events such as sports games and concerts.[1]

On the other end of the satisfaction spectrum were activities like standing in line, lying sick in bed, doing household chores – and scrolling on social media. Yes, that is the company social media was in. Based on this study, you'd think people would bend over backwards to avoid it like trying to get out of doing the dishes. But that's hardly the case.

It's soon been two decades since Apple launched the first iPhone and gave humanity the option to be online pretty much constantly. Today, nearly 80 per cent of the world's population owns a smartphone, which is more than the number of people with access to a functional toilet.[2] [3] If you look around our cities, it's clear that these smartphones aren't exactly inactive. Phone makers like Apple know exactly how much time we spend glued to our screen but it is not public knowledge. The best we can do is look at estimates from outside sources. These suggest an average of five hours a day per person.[4] [5] Numbers vary from country to country, but not as much as you might imagine. Surprisingly, the countries with the highest average screen times don't seem to be high-income countries such as those of the Anglosphere. Rather, it is middle- and low-income countries like the Philippines, Brazil and South Africa. Part of this is due to these countries having populations that skew younger on average. Birth rates are typically higher in middle- and low-income countries, so a larger portion of the population consists of children and young people. And young people spend more time on their phones.[6] The numbers look approximately like this by age group:

- 16–24: 7.5 hours
- 25–34: 7 hours
- 34–44: 6.5 hours
- 45–54: 6 hours
- 55–64: 5.5 hours[7]

Some of this screen time happens behind a computer during work hours, so it is not all social media and the like. But about half of all screen time is on our phones, and social media is the number one cause of this.

*

In the 1950s, the American psychologist B.F. Skinner conducted a few groundbreaking experiments on animal behaviour.[8]

Skinner's goal was to investigate how the brain's reward system functions. In his experiments, he built a few small cages that we now call 'Skinner boxes'. You might recognise his design from the experiments with rats and electrical self-stimulation. In a Skinner box, there is a small button or knob that will release a treat if the lab animal presses it. Again, both mice and rats are naturally curious. If you put one into an unknown environment – such as the Skinner box – it will explore its surroundings by sniffing and feeling around. Before long, the animal will coincidentally press the little button and a treat will surprisingly – to the animal – be released. This doesn't have to happen more than a few times before the animal learns the connection. Then, it will start pressing the button on its own to get treats.

Skinner, however, wasn't just interested in teaching his lab animals the relationship between the button and the treats. He also wanted to figure out how animals react to different reward *patterns*. To do so, he conducted variations of his experiment in which the treats weren't released every time the lab animals pushed their little button.

Naïvely, we might have guessed that the rodents would be most attracted to the button when they were sure something good would come from pressing it. But surprisingly, this isn't the case. In fact, it turns out that animals become more engaged when using a varying, unpredictable pattern. Sometimes a press of the button might release a treat. Sometimes it doesn't. This unpredictable reward pattern makes animals *hooked*. They start showing compulsive behaviours, abandoning other activities such as socialising to instead press the button again and again in the hope of getting the treat. If you think back to the chapter on dopamine, it's understandable why this happens. Remember that dopamine is released right after an *unexpected* reward and then strengthens the behavioural pattern that immediately preceded it. This mechanism ensures that we learn how to get our hands on whatever the reward is. But with a completely random reward system, it is impossible to predict when the reward will come. Thus, the release of dopamine doesn't stop. Instead, the behaviour that occurs immediately prior to the possible reward – pushing on the button – is strengthened again and again and again.

Maybe this whole setup rings a bell with you. After all, I used the word 'jackpot' before and this is because casinos prey on precisely this phenomenon. A slot machine is really just a Skinner box for humans. The only difference is that the reward is something abstract in the form of money. But the behaviour of a compulsive gambler behind a slot machine is exactly like that observed in Skinner's rodents. In both cases, the result is the same: a zombie-like state in which the subjects repeat the same motion for hours on end.

In any other context, people would be bored to death by sitting for hours and pulling a lever or pushing a button. Imagine, for example, that you and I go to a casino and offer some of the gamblers a job at our factory instead. Here, they would perform the same movement as they do behind the slot machine: repeatedly pulling a lever or pressing a button. However, we would pay them to do so, meaning the economic return from this activity would be positive. At a slot machine, the odds are stacked against the gambler (that's why casinos try to keep users playing as long as possible) and thus the expected economic return is negative. Play long enough, and it is sure to cost you money. But of course, even if the factory job would pay out more than gambling, most gamblers would reject our job offer immediately. The physical action of pulling a lever might be completely the same but the variable reward pattern makes gambling immersive for the gambler, even if the end result is ultimately negative. That's how powerful unpredictable reward patterns are. The gambling industry makes immense fortunes exploiting this psychological weakness of ours. But they are actually nothing but a tiny player compared to the industry that *really* profits off B.F. Skinner's discoveries.

★

It's no secret that the creators of social media platforms were inspired by Skinner and other psychologists. Some of them openly admit to it. For instance, in an interview with Mike Allen at an Axios event in Philadelphia, Facebook's first president, Sean Parker, candidly describes the early days of social media like this:

> The thought process that went into building these applications, Facebook being the first of them . . . was all about: 'How do we consume as much of your time and conscious attention as possible?' And that means that we need to sort of give you a little dopamine hit every once in a while, because someone liked or commented on a photo or a post or whatever. And that's going to get you to contribute more content, and that's going to get you . . . more likes and comments . . . It's a social-validation feedback loop . . .

> exactly the kind of thing that a hacker like myself would come up with, because you're exploiting a vulnerability in human psychology. The inventors, creators – it's me, it's Mark [Zuckerberg], it's Kevin Systrom of Instagram, it's all of these people – understood this consciously. And we did it anyway.[9]

Many influential people in the early days of social media companies learned about Skinner's discoveries as part of their university education. At Stanford University in the heart of Silicon Valley, there is a course called Persuasive Technology, in which Skinner's experiments with mice and rats are a key part of the syllabus.[10]

Since these days, it has been entrenched in the culture of the companies and helped make them the richest and most powerful companies in the history of the world. In turn, all this money allows big tech firms to hire armies of world-class engineers with the same goal described by Parker: to get you to spend as much time as possible on social media.

The reason social media has this overarching goal is because their revenue comes from ads. So, the calculation is simple really: the more time you spend on Facebook, Instagram, TikTok, LinkedIn and so on, the more ads they can show you and the more money they make. While you're online the money keeps rolling in, but as soon as you close the app, the cash flow stops. This means that there is a one-billion-dollar question: how do you get someone to look at a screen for as long as possible? Let's take a look.

*

One of the first Skinner-inspired innovations of social media was the 'like' function. Here, social recognition is dosed out to the user, varying from post to post in a somewhat unpredictable pattern.

A study by Karthik Srinivasan from the University of Chicago has showed how this dosing of social recognition affects our behaviour. Srinivasan looked at how TikTok and Reddit users behaved after having a post get an unusual amount of attention. That is, how users respond to hitting the social recognition jackpot. He found that this viral success led users to significantly increase their posting frequency later.[11] The algorithms on social media likely exploit this jackpot-like effect by ensuring that posts get lots of exposure right after we've created an account or if we post something after having been inactive for a while. In both cases, showing more people our post will increase the probability it will get many likes and subsequently increase our usage.

It's hard not to admire the cleverness of this design a little. Because not only does it hook the user, it also automatically makes them provide free work for the social media platform. We quickly learn that not all content results in an equal reception, so we start striving to make exciting content which will in turn help keep people's attention. If you are another type of media – say a newspaper or a magazine – you have to pay people to do work like writing or taking photos for you. But social media gets around this by paying in recognition, and thus saves itself a fortune.

Ultimately, the efforts of creators mean that the content itself is a greater draw for many than the possibility of receiving social recognition. Most people who use social media actually rarely post, but primarily spend their time scrolling and consuming the content. The rule of thumb is that between 1 and 10 per cent of users are responsible for most online content. The rest are largely passive.

You could say social media platforms include two social 'classes' that the company needs to manage: a creator class and a user class. As the user class is so much bigger, it is responsible for the majority of the ad revenue. So it is important to keep them engaged. The companies have optimised this in several ways, one of which is a sort of digital version of the Buffet Effect from earlier. Remember? The mechanism that makes us overeat at a buffet because variety delays satiety. A similar principle applies to social media content. First there's a cat video. Then a political post (which you agree with). Then a funny story. Then a news update. And so on. You might get satiated with each topic if it was the only thing shown. But by serving you a digital buffet, social media platforms delay the feeling of boredom and keep you engaged for longer. That's not always how it's been but a result of optimisation for keeping users around. You might remember in the beginning of social media that you were only shown content from your friends or from organisations that you had chosen to follow. That was part of the sales pitch: come follow along in the lives of your friends! But it has since turned out that Hannah from high school and John from work aren't necessarily the best at creating content that increases your screen time. So, in the quest to keep you engaged, social media platforms transitioned to the buffet model. On Instagram, you don't just see pictures from people you know anymore, but also videos and pictures from popular accounts you don't follow. On LinkedIn, the posts in your feed aren't just from your connections, but also popular posts from accounts you're not connected to. And on TikTok the algorithm practically does all of the work, no user preferences required.

These apps have discovered that you don't always choose the content that keeps your attention the longest. So instead, they make the choice for you. To do this, they collect enormous amounts of data about you while you're scrolling.

Some people are opposed to this data collection and are demanding limits to protect their private lives. But many shrug their shoulders. If you have nothing much to hide, who cares what some company knows about you?

Social media platforms, however, aren't gathering data to find out where you live so that they can steal your grandmother's old silver spoons. They are collecting data about you because it helps them crack the code on how to keep your attention. The second you open the app, the analysis begins: how long did you look at post A? What about post B? Did you click on this picture or not? How long did you stay on the app? What was the last thing you saw before you closed it? And so on and so on.

All this data is fed to a machine-learning algorithm that is tasked with figuring out how to keep you behind the screen. If you scroll quickly past one post or shut the app, the algorithm learns that it should show you less of that type of post in the future. If you pause on a post, 'like' it or share it with a friend, it learns to show you similar things in the future. Hence, as you use social media, you are teaching it precisely how to capture your attention in particular.

This information can be supplemented by information the app has on other users like you. The algorithm can sift through millions of people to find others that behave in the same way as you do. Their behaviour can then be used to predict your reaction to posts beforehand. If, for example, we both react similarly to different content, the algorithm can use my behaviour to decide which posts you should see. If it registers that I 'liked' a video of a funny dog, it will show you the same video next time you open the app. Similarly, if I quickly scroll past a post about astrology, the algorithm will avoid showing you a post about how the new moon marks the beginning of an introverted period of your life.

*

Once social media has got you hooked, it's dead set on keeping your attention.

That is why social media platforms are designed to make your scrolling experience as frictionless as possible. Every pause is an opportunity for you to reflect and remember that you had planned on doing something else. It is essential to keep you in the loop and minimise distractions.

One of the ways social media companies make your experience frictionless is by using so-called infinite scrolling. You might remember that you used to be able to actually reach the bottom of a page and then you'd have to manually go to the next page or refresh the site. But with endless scrolling, updates are continuously loaded so you never need to do anything. Instead, you are presented with an endless feed that never runs out of posts for you to look at. The infinite scroll works so well that Aza Raskin, who invented it, has made a public apology to the world. Among other things, he says 'I regret not thinking more about how this thing would be used.'[12]

Raskin has since helped launch an organisation called the Center for Humane Technology, in Silicon Valley. This organisation attempts to inform people about the manipulative practices of social media companies and their effects on us. One of the other founders is Tristan Harris, who you might know from the 2020 documentary *The Social Dilemma*, in which a group of engineers and entrepreneurs from Silicon Valley go to war with the monster they helped to create.

In general, it is striking how many people from the early days of social media are now fighting it.

Chamath Palihapitiya, a former Facebook executive, says he feels 'tremendous guilt' over his work on the social media.[13]

Tim Kendall, who helped establish Facebook's ad-based business model now runs an app called Moment, which helps people *limit* their social media time.

Roger McNamee, who was an early Facebook investor, has written a book with the subtitle *Waking Up to the Facebook Catastrophe*.[14]

Justin Rosenstein who helped create the Like button says: 'I find myself getting addicted – yes, in some cases, by the very things I've built.'[15]

This story also plays out among many of the mid-level employees in Silicon Valley. The *New York Times* has reported that it is widespread for Silicon Valley nannies to be given strict instructions on keeping kids away from the very screens their parents work to hook other people to.[16] [17] I'd say it isn't always the best sign when those who spend their work hours developing a product fight tooth and nail to keep their own kids from being exposed to it.

In their defence, few of the people working on this stuff 10 or 15 years ago could have anticipated what it would result in. It's one thing to develop a little app and use a few psychological tricks you learned at university to get things up and running. And something entirely different to be able to suddenly control and direct the attention of billions of people. At the very least, we should be grateful that some insiders are leading the fight to stem the negative effects of the products they helped create. Because, ironically, they're often the best people to advise us on how to cut down our screen time. For example, Tristan Harris has offered an alternative piece of advice, which is one of the most impactful I've encountered so far.

However, it requires a detour of a few million years.

*

Like many good stories, this one begins in the time of the dinosaurs. This was when the earliest mammals lived: small squirrel-like creatures that were mostly nocturnal. We can't blame them for that, given their giant carnivorous neighbours.

At night, early mammals didn't need to see colours and they were basically colour-blind. Instead, they used their sense of smell and hearing to orient themselves so that they could find food and avoid predators.

Though millions of years have passed, we see the legacy of this trait in many mammals today. Most have something called dichromatic colour vision. This isn't the same as seeing in black-and-white, but it means that these creatures inhabit a less colourful world than you and I. Deer and wild boar, for example, have dichromatic vision and therefore can't tell the difference between green and orange. As a result, one of their largest predators – tigers – are orange. In the eyes of their prey, that makes the big cat perfectly camouflaged against a green background because the two colours look identical to them.[18]

We humans don't have dichromatic but *tri*chromatic colour vision. Di- and tri- refer to the number of a certain type of cells called cones in the retina of the eye. These cells capture colours, and we have three versions of them. The first version captures bluish colours, the second captures greenish colours and the third captures reddish colours. The cone for red colours is what most other mammals lack and what makes us able to distinguish between green and orange.

We share our good colour vision with other primates. And this is probably for the same reason as our sweet tooth. The earliest primates were fruit-eaters in the rainforest, where bright colours are a sign that fruit is ripe and can be eaten.

A secondary effect of this instinct, however, is that primates are drawn to bright colours. As with our other instincts, this one hasn't escaped the superstimuli machine of social media. Just look at your phone screen. It is beaming with bright and saturated colours that are sharper and more captivating than the colours of the real world. (And we have already decorated the real world by painting it all kinds of colours to please our senses.)

The power of bright colours is especially clear if you set your screen to black-and-white. Because it doesn't take much more than that to make your phone boring. In fact, it can almost be depressing to look at a colourless screen for too long.

### Colourful birds

Humans are better at seeing colours than many other mammals, but compared to birds we're basically colour-blind.

Most birds have *tetra*chromatic vision. This means that they have four types of cone cells, one more than us and two more than most mammals. You might consider that next time you hear birds chirping outside your window. Maybe they're tittering about how terribly colour-coordinated your outfit is because you can't see the colour nuances they can.

Some birds have to camouflage themselves in shades of grey and brown to avoid predators but due to their powerful colour vision others are immensely colourful to impress their conspecifics. Just think about the tail of a male peacock.

The tail doesn't have any physical function but is exclusively made to impress female peacocks with its pretty colours. The message is something like: 'See how healthy and fit I am. I'm so superior that I can please you with my beautiful colours instead of hiding from predators like a coward.'[19]

When all goes according to plan, the females are impressed and mate with the male. But sometimes, the tail catches a tiger's attention, and the peacock becomes an aesthetically pleasing dinner instead.

★

What starts as vanilla ice cream nearly always becomes vanilla ice cream with peanut butter, chocolate chips, marshmallows and caramel sauce. The same is true of social media. So far, we've covered the first escalations: the 'like' function, the infinite scroll, and the transition from friend-generated content to a digital buffet.

This evolution is still under way, and the latest escalation has been introduced by the Chinese-owned app TikTok.

Since the end of the 2010s, TikTok has been wiping the floor with all the other social media. Basically, they have done so by discovering an even more effective way to harvest attention than the competitors. TikTok has been so successful that other social media platforms have now copied their design in the form of Instagram Reels, Facebook Reels, YouTube Shorts and Snapchat Spotlight in an effort to keep up.

The secret behind TikTok's success is that they have taken the superstimuli machine to the next level on multiple fronts at once. The content of the app is ultra-condensed into short video clips, often with music and a variety of added effects. The music isn't entire songs, just the best parts with the less catchy pieces clipped out.

Instead of choosing a video yourself, the app automatically plays one as soon as you open it. That removes the marginal friction that can arise from needing to decide. You might remember from earlier that having many options can make it difficult for us to choose. This problem has afflicted other video apps such as Netflix and YouTube. Due to choice overload, you might end up closing the app without watching anything. But by automatically playing a video TikTok deters the user from quitting the app immediately. The first video shown might not be a hit, but the user can easily swipe on to the next one. At no point does TikTok ask anything of you to display content. Everything is decided by the algorithm and presented in an infinite feed.

This strategy has also been successful because the video-based design means that you only see a single piece of content at a time. On traditional social media platforms, there are usually multiple posts on the screen simultaneously and this makes it more difficult to track what is capturing your attention. Is it the political discussion at the top of the page? Your colleague's holiday photos beneath that? Or maybe the banner ad for a new kind of dental floss? TikTok sleekly gets around that uncertainty, which helps the machine-learning algorithm to decode your preferences even more accurately.

So what we've learned from the success of TikTok is that the easier it is for the machine-learning algorithm to study you, the better it will become at capturing your attention. This is a problem because machine-learning keeps getting better. And also because social media apps don't distinguish between what captures our attention and what is actually beneficial or healthy for us. In fact, one of the lessons from social media is that happy users aren't necessarily the most engaged. Anxious or angry users can also be a goldmine. If you watch a post that makes you angry, there's a relatively good chance that you will interact with it.[20] You might pick a fight with the owner of the post in the comments, or you might share it with your friends to confirm how stupid people are. What social media registers, however, is that *this* content caught your attention. And so, it will show you more from the same drawer.

# 15

# A little fish in a big pond

In the summer of 2012, Michael Phelps became the most decorated Olympian ever when he won gold in the 4x200-metre freestyle relay with the American team. The victory pushed Phelps past Soviet gymnast Larisa Latynina who won 18 medals between 1956 and 1964. By the end of the London Olympics, Phelps had reached 22 Olympic medals, and four years later in Rio de Janeiro, he built on his record and retired with a total of 28 Olympic medals, 23 of which were gold.[1] This is ten more medals than the second-place holder and 12 more than the second most decorated swimmers.

So if I were to describe the swimming abilities of Michael Phelps in a Danish temperament, I'd say he is pretty good. If you wanted to be a little more dramatic about it, you might ask whether Phelps was actually made for swimming fast. Because even if you tried, you'd have a hard time designing a human body better suited to swimming.

First, we know that it helps to be tall if you want to swim fast.[2] Tall swimmers technically have to swim slightly less distance in a given race than short swimmers because the lane has a set length. But besides that, studies show that tall swimmers experience less water resistance at high speeds and can therefore swim faster. Michael Phelps is 193 centimetres tall, so he is taller than 95 per cent of American men and even more than that on a global level.

Next, it is also advantageous for a swimmer to have long arms.[3] Again, this makes the distance the swimmer has to cover slightly shorter, but long arms also help displace more water with each stroke. So the longer your arms are in relation to your body, the better propulsion you will have. Arm length is often measured as so-called wingspan, which is the length from fingertip to fingertip when you have outstretched arms. Usually, a person's wingspan is about the same as their height. That is, a person who is 170 centimetres tall would usually have a wingspan of around 170 centimetres. This rule, however, doesn't apply to Michael Phelps. Even though he is 193 centimetres tall, his arms are so long that he has a wingspan of 203 centimetres[4] – that of a person 10 centimetres taller than him.

As opposed to long arms, long *legs* are a disadvantage in swimming because long legs increase water resistance more than they help you move forward. And once again Phelps has an optimal build. He might be tall, but he only wears size 32 trousers, which would be normal for a man who is 10 centimetres shorter than him.

Finally, it is also important for swimmers to have large hands and feet because this helps you move as much water as possible. Just think of how much faster you can swim wearing flippers. With a European shoe size of 48, you could say that Phelps has in-built flippers, and on top of that, he has hypermobile joints, which gives his strokes extra range.

In fact, Phelps's physique is so atypical that as a child, doctors worried he might suffer from a genetic disease called Marfan's syndrome, the symptoms of which include hypermobility in the joints and an elongated body. But nothing is wrong with

Phelps. He just happened to be born with a body that is abnormally suited to swimming fast.

Of course, you don't reach the top of a global sport by physique alone – it also takes major mental fortitude. But even in this area, Michael Phelps is atypical. He recalls that he was hyperactive as a child but learned to channel his enormous energy into the swimming pool when he started swimming. Afterwards, he became a training fanatic and, leading up to the 2004 Athens Olympics, he recalls training every single day for five years in a row even on his birthday and on Christmas Day.[5]

So, I guess we can conclude that Michael Phelps isn't exactly the most typical human. But while all these traits are remarkable, they are much more unusual when you compare Phelps with the general population than if you compare him to other elite swimmers. You simply don't reach that level without having at least the majority of the 'swimmer's package'.

This is true of all global sports today. The elite athletes are not representative of the general population. They are a subset of atypical humans whose characteristics happen to help them perform at the highest possible level. Just think of the classic example, basketball. The average player in the American NBA is 2 metres tall, putting them in the top 1 per cent of Western men. The best basketball players tend to be even taller than this. And when players are relatively short (for a basketball player), they tend to make up for it by having really long arms. This means they aren't necessarily 'short' with their arms above their heads which is what mainly matters because this is where the ball is released and where most blocks are made.

The opposite example is a sport like gymnastics. Here, it is an advantage to be small because it makes rotating in the air and balancing easier. Female Olympic gymnasts are only 155 centimetres on average, and the world's best gymnast, Simone Biles, is even shorter. At 142 centimetres she is shorter than 99.7 per cent of the Western population and so short that she falls within the limit of what is called dwarfism.

The sports world is the easiest place to see how atypical elite performers tend to be because many of the differences are physical. But this phenomenon is not confined to sports. It's true of every discipline where the arena is global. In science, for instance, Hungarian-American physicist John von Neumann is often considered one of the most impactful scientists of all time. He might have looked like a pretty normal guy – physically – but his brain and cognitive abilities definitely weren't. At the age of six, when most kids are learning the alphabet or how to tie their shoes, von Neumann was able to divide eight-digit numbers in his head and had taught himself ancient Greek, Latin, as well as five or six other languages.

*

One of life's great questions is whether it's better to be a small fish in a big pond or a big fish in a small pond.

In other words, is it better to be a mediocre player in one of the big football clubs or to be the best player on your local team? Is it better to be an average student in a school for gifted children or the star student in your standard school class? Is it better to be overlooked at the hottest nightclub in London or to be the most beautiful person at the local pub?

Your answer probably depends on what pond we're talking about and on what you're trying to achieve. But from a psychological perspective the answer is surprisingly simple: be a big fish in a small pond. Then you'll be happiest.

This phenomenon can be seen at university science faculties around the world. In most countries, there is a goal to get as many young people as possible to study STEM subjects – science, technology, engineering and mathematics. So, sociologists want to figure out how we can push young people in that direction and whether we can predict who will end up actually graduating. Unsurprisingly, the research has shown that it helps to be good at mathematics, if you want to complete a university education in the natural sciences. Take the American university Hartwick in the state of New York.[6] At this university, the students with the strongest mathematics skills make up the majority of those who graduate in scientific fields. If we divide incoming students into three groups according to their mathematical abilities, it looks like this:

- The top third in mathematical ability → 50 per cent of science graduates.
- The middle third in mathematical ability → 27 per cent of science graduates.
- The bottom third in mathematical ability → 18 per cent of science graduates.

These numbers might seem a little clunky to grasp, but the point simply is that many of those who end up graduating in the sciences come from the group of students that is best at maths. This pattern holds true across American universities, even at Harvard, which is probably the most prestigious university in the world.

But here's the catch: Harvard students are in a league of their own. A 'bottom' Harvard student is likely to be better at maths than a 'top' Hartwick student. But that is not reflected in their chances of graduating in the sciences. Here, the mathematically weaker 'top' Hartwick students have a better chance than the mathematically stronger 'bottom' Harvard students.

How can this be? Those 'bottom' Harvard students are unusually talented. But not necessarily so in their own eyes. You see, the reason for this distortion is a psychological phenomenon called the 'big-fish-little-pond' effect.[7] [8]

This effect describes how students with academically strong peers judge their own abilities to be *worse* than they actually are. We can understand the phenomenon with an example. Imagine that we gather two groups of students that are equally intelligent. We send one of the groups to a school for gifted children. The other group, we send to a normal school where the average level is significantly lower. After a few months at these different schools, we ask our students to evaluate their own abilities.

We know that the students are equally intelligent so if their self-understanding reflects their abilities, they should judge themselves to be of equal intelligence. But instead, the ones who are small fish in a big pond – at the school for gifted children – will judge their abilities to be worse than the big fish in the small pond at the normal school.

We say that students with academically strong peers develop a poor 'academic self-concept'. This is important because your academic self-concept influences how much you achieve. If you believe in your own abilities, you are more willing to put in extra effort and not give up when you are presented with complex problems. Self-confidence begets ambition and without ambition you will never shoot as high.

So academically strong students might do themselves a

disservice by choosing the most prestigious universities, if that means they end up as below average. Because among their peers you'll find some of the world's most gifted individuals. We could call them the Michael Phelpses of university students. Even the 'worst' students at Harvard are still in the top 0.1 per cent globally, but their lived experienced will not tell them that.

Humans are prone to a cognitive bias that the psychologist Daniel Kahneman call 'what you see is all there is'. We tend to craft our image of the world based on the information that is right in front of us. This is why students at elite universities can struggle to see their own abilities objectively. The 'bad' students at Harvard might even logically understand that they are enormously intelligent in the global scheme of things. But this won't be what they instinctively feel.

This is the essence of the 'big-fish-little-pond' effect. Sociologists haven't only observed it among American students, but also across all cultures studied.[9] It has been observed in multiple Western countries, but also in non-Western countries like Indonesia, Thailand and Tunisia. And it appears across age groups and in both girls and boys.

The 'big-fish-little-pond' effect isn't limited to *academic* self-concept either.

For instance, in the sports world, athletes also judge themselves in comparison to how good their teammates are.[10] [11] The reason for this universality is that the 'big-fish-little-pond' effect springs from one of the deepest elements in our social nature: our desire for social status.

★

Sociologists distinguish between two forms of social status: the kind that comes from dominance and the kind that comes from prestige.[12] Dominance is the most common form of social status in the animal world and is based on violence: do what I say or else I'll beat you.

We can see this form of social status among humans as well. Consider a gang of criminals, for instance. People don't listen to the Mafia because they like them or because they want to help the Godfather earn money. They help because 'it would be unfortunate if something were to happen to you . . .' In a criminal group, power is roughly distributed according to a person's perceived ability to exert violence against other people and win, either by their own hand or with the help of their cronies and allies.

In a milder context, we also see dominance-based hierarchies that aren't based on physical violence but still at least partly based on the ability to inflict harm. For example, your boss has dominance over you. Not because they will beat you up if you don't listen, but because they can fire you and thus harm you by robbing you of your income.

When all is said and done, however, dominance-based status isn't the primary form of status among humans. We are special like that. Even among our closest relatives, chimpanzees, the social hierarchy is primarily based on dominance. But among humans everything changed once our ancestors invented weapons.

You see, it is not so easy to dominate an angry person with a spear. Chimpanzees take a long time to kill each other by biting and hitting but an armed human being can kill another person in an instant. That makes it a lot harder to be a bully. Who knows – if you terrorise someone from your tribe too much, they might suddenly appear next to your grass bed in the middle of the night, ready to strike.

Besides weapons, humans also moved away from solely dominance-based status hierarchies, when we invented language. Talking to each other allows the exchange of information but also the making of plans. This means that multiple tribespeople can coordinate and gang up on and kill a bully, even if the bully is larger and stronger than each individual member of the coalition.

The anthropologist Richard Wrangham relays a story about how something like this played out among a tribe of hunter-gatherers called the Yanomami, who live deep in the Amazon rainforest. A man in the tribe had been aggravating the other members for a long time by being arrogant, aggressive and generally acting like a bully. One day, when the men of the tribe were out on a hunting trip, they challenged the bully to climb a tree and collect some honey. The bully took them up on the challenge and put down his weapons to do the climb. While he was in the tree, the others calmly gathered his weapons and once the bully returned to the ground, they promptly murdered him and went home.[13]

In less dramatic cases, hunter-gatherers and other small groups of people can force bullies to behave through gossip and social exclusion. Together, this means that most hunter-gatherers have egalitarian cultures where people are careful not to get on each other's nerves and try to treat other tribe-members well. The transition to agriculture and later to complex societies has once again made it easier to exert dominance over others. For instance, a king doesn't have to worry about a single or a few peasants disapproving of him. But generally, dominance is still not the most common form of human social status. Rather, it's the other form of status: *prestige*. Unlike dominance, we don't see prestige status in many other species. But as so often in this book, there is a small bird that can teach us how this form of status works.

This bird is called the Arabian babbler and it lives in small, consolidated groups of up to 20 birds. Each group shares a territory that usually consists of a tree and a few bushes. Here, they have a common nest, where the group's young are hatched, and they cooperate on defending their territory against predators and against other flocks of babblers. This lifestyle makes the Arabian babbler unusual among bird species, but also makes them more like us.

You see, both an Arabian babbler and a Stone Age human need their flocks to survive. Technically, a hunter-gatherer can survive on their own if they are smart and physically healthy. But a single human is an easy victim. We are physically weak compared to most animals and are also vulnerable to other humans on our own. A solitary Stone Age person would have to be lucky not to be robbed, taken as a slave or killed. Similarly, a lone Arabian babbler doesn't stand a chance against a whole flock. If a babbler is on its own, it will have to settle for territories that none of the other birds care for. Probably somewhere exposed, where there is less food and a greater risk of being preyed on.

So, the Arabian babblers instinctively know, as do we, that they're better off as part of a flock. And we can see this in a funny type of behaviour. Arabian babblers don't just collaborate. They *compete* to be as helpful as physically possible. Even among social animals this is rare. In a pride of lions, for instance, the male is the most dominant one and uses his position to eat first after which the females will sometimes fight each other to get the lion's share. Among Arabian babblers, however, the opposite happens. These birds *give* food to each other. It isn't unusual to see a poor inferior babbler stuffed full, and desperately trying to escape a dominant bird trying to force feed it.

The same pattern appears in other behaviours that are beneficial to the flock. Arabian babblers compete to stand guard over the flock, even though this makes them more exposed and limits the time they have to find food. And they fight to take the most vulnerable sleeping position – at the outermost edge of the group's shared sleeping branch – even though this is where they have the greatest risk of being eaten by a predator.

One would hope that the explanation for this type of behaviour was pure goodness of heart, but unfortunately this isn't how evolution works. Arabian babblers aren't aggressively helpful because they are selfless but because it improves their own chances of survival. By making themselves useful to the group, the babblers reduce their risk of being kicked out. It is in the best interest of the group to have members that gather a lot of food or are good at protecting the group.

So, an Arabian babbler who doesn't help out might be safer at night and not have to worry about food. But this bird is easier to do without. Fewer will miss it, if social politics means it is thrown out of the group one day. And once alone, it *really* needs to worry about food and safety. As a result, an Arabian babbler instinctively wants to put its life on the line for the benefit of the flock.

You might recognise some of this bird's behaviour. It is, for example, not random that the captain, who is highest in the hierarchy of a ship, is the last to leave a sinking ship. It also used to be customary for a king or general to lead their troops on the front line. And today, the coach and star players of a sports team often take the heat if the team underperforms.

For us humans, however, competing for prestige isn't solely about *doing* something beneficial for 'the flock'. The competition can also be a matter of being *able* to do something that could be of benefit. For example, athletes have high status in our society. At the first glance, it might seem strange that Cristiano Ronaldo or LeBron James are some of the most admired people in the world, but from an evolutionary perspective it makes sense. Athletic abilities are useful for a group of hunter-gatherers. If you are strong and fast, you can become a good warrior and a good hunter and therefore you are a good ally for the other members of the tribe. So, when people perform athletically, we notice. Especially if there is an element of competition, so the athlete isn't merely demonstrating their impressive abilities but also show that their ability is higher than that of others.

Similarly, we often see that people with many resources – in present-day society this mostly takes the form of money – have higher social status. Again, such a person is a good ally, exactly like an Arabian babbler that has found a lot of food.

In many cases, prestige-based hierarchies are more pleasant for their members to take part in than ones built on violence. But they are not without their own problems. After all, they are still hierarchies and we can't all be more athletic, well-off or intelligent than the average.

This is where some of our less flattering sides come out, most noticeably our tendency to compare ourselves with others. After millions of years of evolution, we instinctively understand that prestige is advantageous. So, we scan our social environment for areas where we can stand out, be useful and get social recognition. When we succeed, it feels good. But when others turn out to be better than us, it doesn't. As with the Arabian babbler on the innermost part of the branch, if we cannot contribute we're a less valuable ally and that makes us vulnerable.

In the small ponds of a village or tribe, this problem can be overcome by most due to the numbers involved. There will not be that *many* people, so most will be able to find a niche to stand out in and feel like they contribute meaningfully. But in the modern world, our prestige instincts can go totally off-course. Increasingly, we participate in huge ponds such as those we have discussed previously. The kinds of 'winner-takes-all' competitions where top performers are one in a thousand or even one in a million.

In the sports world this transition occurred with the advent of television. Suddenly, people went from following their local football team to having free access to watching teams from around the world. This meant that attention went from being distributed locally among many average athletes to being concentrated at the top among a few players in the absolute elite.

The same is true of disciplines like acting, music and writing where most of the attention also goes to the very best, and competition is fierce.

Social media re-enters our story here because it is the latest example of technology expanding ponds that were previously small.

As we've discussed, most of the content in our social media feeds used to be made by our family and friends, just like people used to follow their local football team. But as social media has increasingly turned to serving us the most attention-grabbing content, the dynamic has changed. Some people are just better at making content than others. On image and video-based platforms like Instagram and TikTok, these people tend to be unusually attractive or entertaining. On text-based platforms like Twitter or LinkedIn they tend to be unusually intelligent, successful or – again – entertaining.

Because we spend hours on social media, the highly unusual people presented by the algorithm become part of our 'reference population'. This in turn messes with our sense of social reality. It gets pushed away from the norm so that we encounter more people who are 'one in a thousand' or 'one in a million' when it comes to beauty, career success, humour or intelligence.

⋆

As mentioned previously, every four years the Danish Health Authorities take surveys of the country's health. This survey is called the National Health Profile and contains a section on mental health, which offers an overview of how Danes are doing as a society.

In the report from 2013, 11 per cent of Danes had poor mental health. In the following report, from 2017, the number had risen to 13 per cent.[14] And in the latest report, from 2021, it's up to 17 per cent.[15]

This alarming development is especially evident among young people. Among 16–24-year-olds, 21 per cent of men and 34 per cent of women have poor mental health. A lot of this attributable to anxiety and depression which has *tripled* among children and young people in Denmark since 2006.

The same story goes for the rest of the developed world.[16]

In Sweden, the numbers of children and young people with poor mental health have doubled in the last ten years.[17] In France, the number of young people with depression has *quadrupled* since 2014.[18] And in the United States 42 per cent of high school students say that they feel persistent hopelessness and grief.[19]

These numbers are alarming enough that they warrant some scepticism. Perhaps we have lowered the bar for giving people diagnoses. Perhaps young people today are simply more comfortable talking about their feelings than previously. Or perhaps this increase just reflects that knowledge and awareness of mental health condition is more widespread today.

These explanations could play a role but they cannot fully account for what is happening. Because it is not just that more young people are *talking* more about their mental health struggles; unfortunately there are also more young people who are *acting* on them. In the United States the number of girls aged 10–14 who have ended up in hospital due to self-harm has tripled since 2001 and the same trend has been seen in many European countries.[20]

Two American psychologists, Jean Twenge and Jonathan Haidt, are among the experts who argue that social media plays a role in this mental health crisis. To understand why we can look back at its development.

In all the affected countries, negative developments start in the beginning of the 2010s, usually 2012 or 2013. Since then, many things have changed in the lives of young people, but the greatest shift has undoubtedly been the rise of smartphones which provide 24/7 access to social media. This shift has changed the social lives of all age groups, but it makes sense that young people have felt it the most. First, because young people are traditionally the age group that socialises the most. And second, because young people have the most screen time, as we have previously established.

A Spanish study has investigated this hypothesis by studying a natural 'experiment'. The researchers exploited the fact that not all young Spanish people received access to social media on their phones at the same time – at least not access of the same quality. On image- and video-based social media like Instagram, the experience is better if you have a good internet connection. So the scientists looked at what happened to the mental health of teenagers when high-speed internet was rolled out in the local area.

The results showed that an increase in hospital admissions for anxiety, depression and self-harm followed for young girls.[21]

American scientists have observed the same pattern as Facebook became part of the daily lives of students at US universities. At first, Facebook was limited to a few universities, but eventually more and more students gained access. Scientists used survey data from the National College Health Assessment to track how students' mental health changed in the subsequent years and, as with the Spanish scientists, they found a decline.[22]

Of course, two studies like this aren't enough to conclude definitively that social media is the reason for the mental health issues of young people today. But it is thought-provoking. And other studies – conducted with different methodologies – tell a similar story.

For example, in an American study, scientists investigated the effects of quitting Facebook entirely. The scientists behind the experiment paid over 2,000 people to participate and divided them into two groups. In one of the groups, participants kept using Facebook as usual, while the participants in the other group were asked to stay off the app for a month. In the end, the ones who quit Facebook were happier, and they also reported less anxiety and fewer symptoms of depression.[23]

A Danish study has found the same thing,[24] but not all studies agree. In some, people don't get happier when they quit social media.[25] But by comparing all the studies that have been conducted to date, Jonathan Haidt and his colleagues have found an interesting pattern.[26]

In short-term studies – conducted over a few days – people who limit their social media use typically don't feel happier. In fact, sometimes they end up feeling *worse*. But over longer periods – a week or more – the positive effects are there. This is the classic pattern of quitting an addiction: withdrawal that makes you feel worse initially followed by improvement in the long term.

★

In people with depression and anxiety you often see what's called 'upward social comparison'. As the name implies, this involves comparing yourself to others who are better off than yourself on one or more parameters: people who are more beautiful, richer, more intelligent or more popular.

As we have established, that's exactly the type of people social media promotes in our feeds – and because of our status instincts, this is a sure-fire way to provoke a feeling of inferiority, the exact feeling common in people with depression and anxiety. For this reason, it isn't so strange that social media can lead to both psychological ailments. As we've already covered, social media warps our image of the world, making it consist of atypical people – those who are most beautiful, most successful and most popular. For that reason, social media makes it easier than ever to make upward social comparisons and to make you feel that you are lacking. Of course, you might wonder why we don't just stop comparing ourselves with others when doing so puts us in a bad mood. But again, the answer is that we are fighting inborn instincts here. It is natural for humans to be preoccupied with high status members of 'the flock'. The same goes for other primates. Remember the monkey-porn study from earlier? Here, scientists were able to get male rhesus macaques to 'pay' with fruit juice in return for seeing pictures of the behinds of female monkeys. But this actually wasn't the only thing the monkeys were willing to pay for. It turns out both sexes where willing to give up their fruit juice in return for looking at pictures of the flock's highest-ranking members.

There are likely several reasons that we are so preoccupied by high status people. First, they have high status for a reason. Or, at least that used to be the case. Today you can become rich and famous for rather trivial things. But in a hunter-gatherer society, high status would most likely mean that you were a top hunter or warrior. Both are beneficial abilities, and even for selfish reasons it would be beneficial to pay attention to people possessing them. Observing their behaviour might reveal clues to the source of their success which we could copy to become high status ourselves.

Another reason we are naturally interested in people with high status is the power they have. Today, powerful people are rather limited compared to previously, at least in the West. But imagine you are a farmer in an early agricultural society. The king of your society *hates* the colour green. So, if you are summoned to see him one day, it would probably be a good idea that you know this and wear another colour. Otherwise, you might never leave the palace . . .

★

The social comparison problems of social media platforms are further exacerbated by the fact that they tend to only present us with the high points of people's lives. Few users post pictures of their double chin while loafing on the sofa. No, there is a steady stream of pictures taken from the right angle, updates about career milestones and posts from exciting holidays. Again, our brains aren't wired to understand that this represents a curated reality. What you see is all there is.

This misunderstanding is evident in studies in which people are asked to evaluate their social lives compared to those of others. These studies show we tend to believe that other people lead more exciting social lives than us even when it isn't actually the case.[27] In theory, social media could be designed to shield us from these unfortunate – and incorrect – comparisons. But that would be bad for engagement, because our status anxiety is the perfect way to ensure people keep making content.

It's not random, for example, that likes, follows and so on are visible to everyone. If the numbers of followers or connections you have were only meant to be a form of reward, it would be sufficient to display them privately. But these numbers aren't meant to be a carrot only. They are also a metaphorical stick. By making them public, social media platforms create a visible status hierarchy. One that many people will instinctively want to climb – both to gain social approval but also to not look unpopular.

Decades ago, we might have known that we weren't exactly the most popular person in school. Some of the other students might have been invited to parties we weren't asked to attend or had lots of friends from other schools. But there was still a certain degree of blissful ignorance. We wouldn't see these the parties happening. And we also wouldn't know our exact position in the pecking order. But both things have changed. Social media platforms let us know what we're missing out on and present us with exact numbers on how popular we are in a format that is directly comparable and visible to others. It begs for social comparison.

So at this point, it isn't so hard to understand the connection between social media usage and mental health issues. It's also easy to see why this problem would be especially pronounced among young people who are trying to find their social standing in the world and who are online the most. One mystery does remain, though. Because if we dive into the numbers and split by sex, both young women and men are more prone to have depression and anxiety than previously. But the problem is *significantly* worse among girls.

In some countries, such as Norway, it is actually exclusive to them.[28] But why might things be going downhill especially fast for young women?

**Negativity is good for business**

An international study investigated how various words in headlines affect the number of people who click on an article. They found that every negative word that could be included increased consumption. The top ten words were: wrong, bad, awful, hate, war, worst, sick, fight, scary and hell.[29]

This negativity bias is an unfortunate part of human psychology, meant to help us survive in our past where danger was constantly lurking around the corner. Now, we have access to information about so many catastrophes, scandals and dangers that it can be completely overwhelming and causes a lot of unnecessary worry and anger. Any sane online environment would be built to dampen fear. Regrettably, there is money to be made here, so that doesn't happen. Social media algorithms don't care what content they need to show you to keep your attention. For instance, in an Australian study, researchers investigated the personal TikTok algorithms for people with and without eating disorders. They found that the algorithms of users with eating disorders send them proportionally more videos about physical appearance, dieting and even serve them pro-eating disorder videos.[30] TikTok have since said that they will investigate the issue.[31]

# 16

# From cosmetics to nuclear weapons

The American barn swallow is a small bird with steel-blue feathers on its back and an orange belly.

As with many other birds, the males are more colourful than the females and the plumage is meant to impress. Female barn swallows are especially preoccupied by the males' breast feathers which can range from light orange to an intense, almost reddish colour. Generally, the females prefer males with relatively darker breast feathers, so these males tend to get more offspring. But scientists have found a way to help the light-chested males not die alone.[1 2]

All you have to do is catch them and paint their breast feathers darker with a cheap marker. Then, boom . . . Casanova status. The females begin to prefer the 'enhanced' males who experience a rise in testosterone and who end up having more offspring. Does this story ring a bell? As humans, we actually use superstimuli to make ourselves more attractive too. But the clearest analogy to the barn swallow among us has the sexes reversed.

This means that we've reached the point of the book when I, as a man, will attempt to write about women's appearance without sounding like a jerk or getting hunted by an angry mob when this book comes out. So, why don't you sit back and have some fun watching me try.

First, it is worth mentioning that it is more difficult to scientifically assess the appearance of humans compared to birds. As humans, we are cultural creatures, so what we find beautiful isn't only biological. It is also a matter of cultural interpretation. In many cultures, beauty ideals are quite different from what they are in present-day Western countries.

In the Mursi and Surma tribes of Ethiopia, for example, women expand their lower lips with clay stones, so they look like a dinner plate. I suspect there might be a die-hard porcelain collector or two out there who would find this appealing. At least in Denmark, the country of 'Royal Copenhagen'. But I'm sure we can agree this trend doesn't exactly align with Western beauty standards.

Other examples like it include the Kayan people of Myanmar, where women wear a collection of rings around their throat to warp their collarbones, making their necks look abnormally long. And then there is ancient Japan, where women used to dye their teeth black to look more beautiful. Neither of those would be especially popular on a dating app in a Western country.

However, even if these examples prove that our beauty ideals are partly cultural, that doesn't actually mean that you cannot make superstimuli for them. You see, in psychology there is a learning phenomenon called the 'peak shift' effect. Psychologists and neuroscientists have observed that when an animal is taught that an object or a pattern is associated with reward, it often develops a preference for an exaggerated version of that object or pattern.

For example, you can teach rats to distinguish between triangles and rectangles. The rats will be shown pictures of each shape side-by-side with a button underneath them. If the rat pushes the button under the rectangle, it gets a treat. Then you repeat until the animal understands that rectangles are good, and it starts choosing them every time.

In these experiments, it turns out rats will react even more strongly to rectangles that are not like those they were trained on but exaggerated versions of them. That is, rectangles that are even more distinct from triangles than previously. This could be a rectangle that is narrower or longer than previously. This might be a little hard to follow, so let's make an imaginary experiment to illustrate the point. In our experiment we show rats two cards. One is light grey and the other is dark grey. When they peck the light grey card, we reward them with food. When they pick the dark grey card, we don't reward them. When the rats have learned this pattern, we suddenly show them the original light grey card and an even lighter one. According to the 'peak shift' effect we would now expect the rat to prefer the card that is even lighter than the one it was trained on. It's as if the rat has not just memorised a specific rule like 'this exact light grey card is good'. Rather, it has learned a relationship between the stimuli 'the lighter the card, the better' and then it follows this rule to its logical conclusion.

The reason this is relevant to us is that the 'peak shift' effect shows that even if our beauty ideals are learned – something that we have arbitrarily decided upon in our culture – it should still be possible to design superstimuli for the underlying rules. This can explain why cultures tend to go to the extremes in their beauty standards. What starts as 'hey, that necklace looks nice' gets exaggerated into 'so many rings around your neck that your collarbone is deformed to accommodate them'.

★

Even though our beauty ideals are partially conditioned by culture, we can't get around the fact that there *is* a biological aspect of beauty, too. There are certain innate preferences that are part of human nature. This is clear in studies in which scientists show pictures of various people to study participants from different cultures across the globe and then ask them to evaluate how beautiful the people in the pictures are.[3] [4]

In these experiments, people from different parts of the world generally agree. Usually, the same people are judged as beautiful no matter whether you ask people from Europe, Africa or Asia to evaluate them. And regardless of whether the people in the pictures are themselves European, African or Asian. In other words, a Canadian person and a Korean person will largely agree on how attractive a given Canadian or Korean person is. This is a result of our underlying and shared human nature.

The findings of cross-cultural agreements on beauty applies to both sexes. But while both sexes are judged by their appearance in Western society, I'm sure we can agree this is particularly pronounced for women. This is the opposite pattern to most animals. Just think of the poor light-chested barn swallows whom females didn't want or the poor male peacocks who have to drag around a giant 'come and eat me sign' in order to get noticed. The reason why women are judged the most on their appearance among humans could again be cultural or biological. But in any case, it is possible to employ a superstimuli. One that is surprisingly similar to the marker trick of male barn swallows: makeup. You see, for barn swallows, the dark breast feathers are what is called a sexually dimorphous trait. That is, a trait that differentiates between the sexes. For other animals this might be a male lion's mane or the antlers of a deer. In both cases, females of the species are attracted to the sexually dimorphous traits and the rule can often be 'the more exaggerated the better'. This fits with both our knowledge of the 'peak shift' effect and superstimuli. So we would guess that if we were tasked with helping improve a lion's chances with the lioness, we should make him a wig or some form of hair extensions that would make his mane look more impressive.

To return to humans, makeup also accentuates sexually dimorphous traits, in the form sexually dimorphous facial features. Male and female faces differ in a number of ways and make-up helps push a woman's face further in a feminine direction. For example, compared to the rest of the face, women's eyes are larger than men's. So, make-up that makes the eyes appear bigger will make a face look more feminine. This is precisely what mascara and artificial lashes are designed to do.

Women also naturally have more colour contrast in their faces than men do. This means that the colour differences between different parts of the faces are greater in women than in men. So, if you take an androgenous face, you can make it look more masculine by decreasing the contrasts and you can make it look more feminine by increasing the colour contrasts.[5]

Again, this is exactly what makeup does. Eyeliner and eyeshadow increase the contrast between the eyes and the rest of the face. Lipstick does the same for the mouth. And blusher can do it for the cheekbones.

*

One early summer morning in 1945, a New Mexico sunrise suddenly became as bright as day. An enormous fire ball illuminated the sky and burned so hot, the desert sand turned to glass. A minute later, a mushroom cloud rose to over 12 kilometres in height.

From a safe distance, a group of scientists and military leaders watched the destruction in silence. They must have felt a mix of pride, fascination and deep existential terror. Everyone who was present knew that they had changed the world forever. Whether for better or worse was still to be determined.

The scientist heading the project, Robert Oppenheimer, described his feelings by citing a verse from the holy Hindu text the Bhagavad Gita: 'Now I am become Death, the destroyer of worlds.' The atomic bomb was born.

Three weeks after the first successful explosion, the new weapon was used in war for the first time. The American bomber *Enola Gay* dropped the bomb 'Little Boy' on the Japanese city of Hiroshima and immediately killed over 70,000 people. At least as many were killed subsequently from injuries, radiation and, later, cancer.

In using the nuclear bomb, the Americans sought to force Japan to surrender and end World War II without tens of thousands of US soldiers having to die while invading the country the old-fashioned way. So, when Japan refused to surrender, the USA dropped another atomic bomb, this time on the city of Nagasaki. That was enough to force a surrender, and so World War II was officially over.

However, the time that followed was not one of peace and rejoicing. With the new bomb the Americans had displayed to the world a form of power the likes of which had never been seen before. They had eradicated an enemy without sending a single soldier into battle. And that made other powers nervous.

Particularly in the Soviet Union, people weren't exactly thrilled about America's newfound might. The two countries had collaborated to defeat their common enemies, the Nazis and the Japanese. But following their victory, it was clear that the two countries would have a tough time getting along. On one side was the communist Soviet Union with centralised economic planning and a single communist party in charge. And on the other side was the capitalist United States with a free-market economy and democracy. The two systems were not compatible and both sides feared that the other would spread their evil ways to the rest of the world.

In the Soviet Union, there were already plans under way to develop an atomic bomb before the US attacked Japan, and the American demonstration only accelerated the development. Some of the physicists in the US nuclear weapons programme sympathised with communism and actively leaked top secret information to Soviet agents, helping them speed up their bomb making. Before long this meant that the Soviet Union was successfully testing its own atomic bomb.

Obviously, it isn't nice to know that your enemy can pulverise you with the push of a button. So, in the decades that followed the two superpowers fought an exhausting battle for supremacy. When the United States invented a nuclear weapon that was twice as powerful as previously, the Soviet Union went to work creating a version that was four times more powerful. Then came the Americans with a bomb that was eight times more powerful, and so it continued. In the end, the most powerful nuclear weapons were 1,600 times more powerful than the two bombs that were dropped on Hiroshima and Nagasaki combined. This is so destructive that if you dropped it over Times Square in Manhattan, the fireball alone would reach into New Jersey to the west, Brooklyn and Queens to the east and past Central Park to the north. And the blast itself would destroy pretty much the entire New York Metropolitan Area, killing millions of people.

As if that is not enough, the battle for military supremacy stretched way past nuclear weapons to regular bombs and missiles, tanks, war ships, fighter jets and so on. As you might expect, this made the entire affair extremely expensive. Instead of spending their money on education, healthcare or fighting poverty, large parts of these nations' wealth were stashed away on military bases with little use.

The leadership of both countries knew this was a stupid use of resources. But how could you get out of the race? If the Americans decided to cut their military spending, it would give the Soviet Union free rein to dictate the world order. And the opposite was also true. What would be the point of saving money if it meant risking being subjugated by your rival?

You might wonder why the two countries couldn't just agree to decrease their military efforts *together*. This was tried but negotiations were delicate. Because what if either country was secretly continuing to stockpile weapons even if they had agreed not to? Then the country that chose to play by the rules could be overtaken by the cheater. De-escalation required trust, and that was hard to establish. It didn't help either that the leadership of both countries was periodically changing. How could each country be sure that the next leader in line wouldn't change course and start stockpiling weapons again?

So, in the absence of a good way to de-escalate, the two superpowers ended up trapped in a military arms race. This phenomenon actually isn't just limited to warfare. It is also prevalent in many other parts of society and in the natural world.

Take, for example, our old friend the peacock. As we know, female peacocks prefer the males with the most elaborate tails. But beauty is relative. For the males it would be nice if it was enough to have a few beautiful tail feathers. Then, they could put the rest of their resources – calories – towards something else. But what if a new male shows up on the block with twice as many colourful tail feathers?

Then he'll get the females. In the next generation, his sons will inherit the extra tail feathers and so it will continue. Clearly, this development cannot last forever – the peacock also needs to survive. But with the tough competition between the birds, everything that can be spared is done away with, and all the resources that can be summoned are spent on decoration. Just like the United States and the Soviet Union ended up spending all their resources on unused weapons.

In both situations the competitors end up in a deadlock. When the Cold War started, the United States and the Soviet Union were roughly equal in military strength. And after decades of nuclear armament – and expenses in the trillions – they were *still* roughly equal in strength. Similarly, male peacocks had nice tail feathers in the beginning of their development. Now their tails have become so gigantic that the poor animals risk their lives to win over females and additionally make it difficult for them to fly. But all of this effort doesn't give anyone much of an advantage. The situation is perfectly described by Lewis Caroll in *Alice in Wonderland*:

> 'In our country,' Alice said, still panting a little, 'you'd generally get to somewhere else – if you run very fast for a long time, as we've been doing.'
>
> 'A slow sort of country!' the Queen said. 'Now, here, you see, it takes all the running you can do to keep in the same place. If you want to get somewhere else, you must run at least twice as fast as that!'

*

We can understand why the nuclear arms race is relevant for our discussion on physical beauty by taking a trip to a parallel universe. Imagine a world in which cosmetics had never been invented. In the real world, make up has been used in various forms thousands of years. Even the Sumerians used primitive versions of eyeshadow and lipstick 5,000 years ago.

But in our fictional world, people have managed to overlook the possibility of making themselves more attractive all the way up to modern times. So everyone just looks like they naturally do. Some people are beautiful, others aren't, and that's just how it is. As in our world, there are some advantages to being physically attractive. Not just when looking for a partner, but also in terms of work, since we know from studies that good-looking people earn more and are evaluated to be more competent than their peers.[6]

In our fictional universe, people are less aware of this connection, but one day, a young woman discovers the advantages of beauty. Let's call her Kimberly. Kimberly realises that she can advance her position in society by improving her appearance. So, she puts on her thinking cap and invents makeup as we know it today.

With her makeup, Kimberly becomes more attractive than she was before, and more attractive than many of the other women. She gets more attention and gets a pay rise. Those benefits are hard to hide, though, and soon enough other women take notice of Kimberly's invention and start doing their own experiments.

In the beginning, only a few succeed, so it doesn't change much for Kimberly. But over time most women in this society are either pulled or pushed into wearing makeup. And so, Kimberly no longer has her advantage. Everyone looks better than they did before. But if we assume makeup enhances appearances equally, they all end up in the same relative position as before. The most attractive people are still the most attractive, and the least attractive are still the least attractive. The only difference is that people now need to wear makeup to keep up with everyone else. If Kimberly is running late one day and *doesn't* put on her makeup, she'll be less attractive than the others and might receive comments such as, 'You look tired today.'

Kimberly doesn't want that. She wants to be the most beautiful woman in the country again, so she puts on her thinking cap. With her cleverness she comes up with a new way to improve her appearance: plastic surgery. She starts with something simple, an injection to make her lips look fuller. For a brief period, everyone is talking about how beautiful Kimberly is with her full lips, but before long the other women have discovered the same procedure.

So, Kimberly takes the next step. Maybe she moves from her face to her body and enhances another sexually dimorphous feature by enlarging her breasts with silicone. Soon enough some of the others follow *that*, and so it continues. Kimberly runs and runs but keeps ending up in the same place.

Okay, this story might be a little too on the nose. But something similar *is* happening in our modern society. It is already standard practice on social media platforms to edit your pictures with apps that make you look better. The most popular one doesn't even try to hide this fact and is called 'Facetune'. On Facetune you can do everything from smoothing out your skin to rearranging your entire face. You can even edit your appearance in videos now.

In the real world, the beauty arms race also means that more and more people are getting plastic surgery. In my native Denmark, there were six times as many breast-enlargement procedures in 2022 as there were in 2005. And there were over 20 times more face- and neck lifts, six times as many liposuction procedures and almost ten times as many eyebrow lifts.[7]

Globally, some countries have arms races that run even hotter than that. In South Korea, for example, plastic surgery is now so normal that 31 per cent of *all* women have had at least one procedure done.

★

Now we're finally ready to answer the question of why so many young women today feel bad and sum up the role of social media. It goes like this:

For cultural and biological reasons, young women tend to be particularly concerned with their appearance.[8] This is not unjustified given that there are demonstrable real-world advantages to physical beauty. However, it is especially troubling in a world dominated by social media. We are instinctively drawn to beautiful people, so given that our attention is valuable, this is what social media algorithms pushes. But beauty is a relative phenomenon and as in so many other instances, we experience desensitisation. So if you only compare yourself to your neighbours, you might feel attractive, but on social media the whole world is part of 'the competition'. It is a really big pond. Those who 'win' in the algorithm and make it into your feed are often the most beautiful among thousands or even millions of people. They are not representative of the population overall, but unfortunately our brains have the cognitive bias 'what you see is all there is'. So, these abnormally attractive people are who we end up comparing ourselves to.

For this reason, a scroll on social media can quickly make people who are prone to 'upwards social comparison' feel bad. We all compare ourselves to others to a certain degree, but young people finding their place in the world are especially vulnerable.

Young people also spend the most time on social media. And young women spend more time there than young men.

In addition, social media makes it plainly visible that physical appearance aids success in life. The beautiful people who dominate the algorithm make tons of money and live luxurious lifestyles, often by doing little more than looking good.

So, it is hard not to be drawn into this arms race yourself in an effort to get the same advantages. But with everyone in the competition, people end up running as fast as they can, while ultimately staying in the same place, only with an increased insecurity about their looks.

# 17

# The ideal body on steroids

Men are naturally just as likely to end up in arms races as women are. The difference is that we judge men's appearance on different parameters. There are also men who get plastic surgery in an effort to look better. But the best example of an appearance arms race for men doesn't have to do with facial features but with *actual* arms. And other muscle groups as well. In studies, heterosexual women typically evaluate men with larger muscles as more attractive, at least up to a certain point.[1] Muscles, as we know, can be enlarged by weightlifting, and this is mostly a win-win situation. You look better on the beach and are healthier at the same time. But there are also artificial shortcuts to increase your muscle growth in the form of what are called anabolic steroids. These are synthetic drugs that mimic the effect of the male sex hormone testosterone.

Testosterone affects the body in a variety of ways, but one of the most visible ones is that it increases muscle growth. This is why men tend to be more muscular than women. Steroid-users often take what is called 'supraphysiological' amounts: amounts that far exceed natural testosterone levels and which in turn cause muscle growth that is impossible to attain in a natural way.

This phenomenon gives us one final opportunity to recall one of the first lessons of this book: the body is dynamic and adaptable. When the body of a steroid-user discovers the huge quantity of testosterone-like substances in the bloodstream, it concludes that the testicles are producing too much testosterone. So, it tries to get levels back to normal by signaling the testicles to decrease their testosterone production.

But since this testosterone-like substance comes from injections, not the testicles, this makes no difference. The testosterone levels stay unnaturally high, and the body continuously reacts by telling the testicles to decrease their production.

Eventually, the users' own testosterone production might have completely ceased. This makes the testicles shrink and can even kill the testosterone-producing tissue.

While the risks are high, the benefits can be, too. It's a well-known secret that many men who make money off their physique take anabolic steroids. This is the case for fitness influencers on Instagram and YouTube as well as action heroes in Hollywood movies. The old guard of action actors are mostly honest about their chemical aides. Arnold Schwarzenegger, for example, has admitted that he took both testosterone and the anabolic steroid Dianabol for years to maintain his formidable physique.[2]

The Swedish actor Dolph Lundgren, who played Ivan Drago in *Rocky IV*, has also admitted to taking steroids throughout large parts of his career.[3] And his opponent in the famous *Rocky* film, Sylvester Stallone, may never have spoken openly about taking steroids. But in 2007 he was stopped at an airport in Australia carrying 48 vials of human growth hormone, which, just like anabolic steroids, increase muscle mass.[4]

Today, fewer Hollywood actors speak openly about steroid use and most credit their physique to working out and eating chicken with broccoli and rice. Of course, they do both of those things as well. But occasionally the names of famous actors, musicians and other celebrities show up in cases where doctors and pharmacists have been arrested for selling steroids.[5] And you have to be pretty naïve to believe 'it's all natural' when confronted with some of the crazier stories from show business. Like when people gain 10–20 kilograms of muscle mass over just a few months. Or when someone becomes significantly more muscular at the age of 50 than they were 20 years earlier.

Of course, it is unfair to accuse everyone of cheating, and just because you use steroids doesn't mean you can skip working hard or a healthy diet if you want an impressive physique. But studies show that people who use steroids and *don't* work out acquire more muscle mass than people who work out without taking chemical aids.[6] So, it's not just a little help. Steroids help build an amount of muscle mass that is simply impossible to build with natural levels of testosterone.

*

It isn't difficult to understand why actors and fitness influencers choose to take steroids. Big muscles are what the viewer wants. And both showbiz and social media are global stages. An army of talented rivals are ready in the wings, tripping over themselves to replace you. Steroids might be that extra push you need to stay ahead.

And so the arms race goes. Musculature is relative and even an amateur bodybuilder from a modern gym would seem insanely muscular if we sent him back 100 years. But on the stage of professional bodybuilders, he would look like he skipped leg day, arm day and every other day. Indeed, professional bodybuilders (or their trainers) now basically need a medical degree to keep track of all the different hormones they take. The competition has become so extreme that some – both professionals and amateurs – are injecting oil into their muscles to make them look bigger. The bar has risen to a level so that even Arnold Schwarzenegger in the best shape of his youth would not be competitive in today's version of the most prestigious bodybuilding competition, Mr Olympia.

We might not all be keeping up with the world of bodybuilders, but alongside fitness influences, Hollywood actors and other celebrities who use steroids, they skew the perception of what is naturally achievable. Just as photoshopped Instagram models with plastic surgery skew our perception of how a normal woman's body looks.

We can get a sense of how this 'arms' race has affected body ideals for men by looking at how children's toys have evolved over time.

In an American study, scientists dusted off editions of action figures like G.I. Joe and Luke Skywalker from past decades. The first versions were from the 1960s and 1970s, when the toys looked like regular fit men. But over the following decades, they became increasingly muscular. Their chests broadened relative to their waists, and their arms got bigger.

The last action figures in the study are from the end of the 1990s. The scientists compared the proportions of these figurines with those of professional American football players to see how unnatural their proportions had become. When they scaled the football players and the action heroes to the same height, the figurines' chests were *twice* as broad, and their biceps were *two to three times* as large. That gives them measurements that not even the biggest professional bodybuilders in the real world could compete with.[7]

Action figures for boys reflect society's body ideals, just as Barbie dolls do for girls. If Barbie was a real person, she would also have some impossible body proportions. Her European shoe size would be a 33 and she would have a BMI of 16. That is so dangerously underweight that it could be life-threatening.

In both cases, it is interesting to note that body ideals are exaggerations of something that starts out as healthy and attractive. It is great to be slim and to be muscular. Only when an arms race pushes these features to the extreme, does it become dangerous. And at this point, it usually also stops being attractive to the opposite sex. Heterosexual men generally prefer women who are relatively thin, but not women with a BMI of 16.[8] And heterosexual women generally prefer muscular men, but usually not bodybuilders.[9] Ironically, it seems that the arms race blinds its most ardent participants such that they overshoot the mark in their quest to stand out.

**Leg-lengthening operations**

Another body ideal for men is being tall. This shares some of the same advantages as being muscular but is not exactly easy to influence. During puberty we grow as new bone tissue is created at the end of what is called long bones. But when puberty ends, the growth plates in these bones permanently close and growth stops.

During the Cold War, the Soviet scientist Gavriil Ilizarov discovered a way to get around this and make people taller. The procedure hasn't changed much since and is pretty brutal. It involves elongating the patient's thighs or shins by breaking and then pulling them apart. First, the bone is cut all the way through. Then, a metal rod is screwed into the bone on either side of the break. This metal rod is used to gradually pull apart the two halves. While this is happening, new bones are formed in the gap when the body tries to repair the break. So over time the operation makes the patient's thigh or shin bones longer, and typically those who undergo the procedure end up becoming 10–15 centimetres taller than they were before. However, the patients need to undergo extensive rehabilitation just to be able to walk again, are at lifelong risk of complications, and have to endure the extreme pain of having their bones cut in half and pushed apart.

Despite the significant pain and risks, the procedures are growing in popularity, though. The BBC has interviewed clinics around the globe, specialising in these leg lengthening procedures, and all report that their patient numbers are growing each year.[10]

*

As you can probably tell, arms races are happening all over the place in modern society. And there is still one final important arms race we need to address. The one that underpins large parts of the book.

You see, throughout *Super Stimulated* we've looked at different industries that exploit our psychological weaknesses to make money – sometimes by going to extreme and unethical lengths. But it's important to clarify that this doesn't mean everyone who works in these industries are evil sociopaths. A few most definitely are. Yet, the majority of the evil we've encountered so far comes from mostly good people who are helplessly trapped in arms races.

Take food superstimuli. In the supermarket, consumers mostly choose what to buy based on what tastes best. So, if you want to make money and beat your competitors, you will be tempted to add extra sugar, fat, salt, flavour enhancers and so on. If not, your competitor will. You might very well be thinking all the right things: 'It's healthier for our customers if they don't eat too much salt.' But if your competition adds extra salt and the customers buy their product instead, what does it matter?

Then your competitor will outcompete you, so your company will have to follow along and raise the salt content of your own product too. Before long the consumers will have grown used to this new, higher salt content. And so, the opportunity arises again: capture market share by adding *even more* salt. Will your company go along this time too? It can be difficult to stop because, without sales, your company will die. That leaves only the competitors who were willing to escalate the arms race. What a mess of a situation. It can seem impossible to quit the race if you're not ready to sacrifice either your company or your principles. So even if the majority of those working in each industry are good people who don't want to hurt others, they risk falling into the trap.

Fortunately, we know of two good ways to stop an arms race.

The first requires involving a powerful third party to set rules for the 'game'. Take the previous example of the United States and the Soviet Union, where the phrase 'arms race' originates. In this case, the United States and the Soviet Union kept building their stockpiles until they were able to destroy the entire planet many times over.

There's no third party in the real world powerful enough to force these two superpowers into anything. But imagine we could call on one. Maybe God, maybe aliens, or maybe Odin and Thor. Whoever it is, they could tell the leaders of the two countries: 'You may produce five nuclear weapons each and spend up to 1 per cent of your yearly GDP on the military. If you exceed these limits, we'll zap all the politicians and turn you into dust.'

In this case neither party needs to worry about being tricked and the arms race will stop. No one needs to fear that the rival will go back on their word, because then the mighty third party will put the rival in their place.

This solution is what we use *within* individual countries. The state plays the role of the powerful third party. They have the monopoly on violence and ensure that citizens don't have to enter arms races to defend themselves against each other (with the US as the unfortunate exception). In an ideal world, the state would similarly de-escalate some of the arms races we have encountered in this book. For instance, by setting various limits for industries that prevent them from competing themselves into extreme territory and hurting their customers. Of course, there will always be the risk of someone breaking the rules and selling illegal products on the black market. Heroin still exists, even though it is illegal. But as we saw earlier with cannabis, legislation and bans help slow the arms races and decrease consumption, even if it's not perfect.

The type of legislation I'm talking about here is the solution China has chosen, particularly when it comes to digital superstimuli. If you open the Chinese version of TikTok (called Douyin), you won't be greeted by silly dances, gossip and sports clips, but rather by short videos about science, mathematics and history. Not because the Chinese are more cultured than the rest of us, but because the Chinese government regulates how tech companies organise their feeds. They aren't allowed to fully optimise their superstimuli but are forced to prioritise certain beneficial topics to an extent. Legislation also requires the app to ask its users regularly if they want to take a break. This is a bad idea if you want to keep hooked to their device and show them as many ads as possible. But it helps the user to break their zombie stupor.

The Chinese, in other words, attempt to slow the arms race by letting legislation intervene in the habits of consumers. They now also require citizens by law to register their real name to play computer games, and playing time is limited for children under 18. They can play for one hour a day maximum and only have access to games on Fridays, weekends and holidays.[11] There is also a debate around whether screen time should be limited to two hours a day for children under 18 and whether access should be blocked between 10 p.m. and 6 a.m.[12]

Of course, rules like this could themselves become dystopian. I'm sure we can both agree it is nicer to live in a society where people are allowed to choose themselves how they want to live their lives.

But even as societies that value freedom, we have already accepted that this unlimited freedom is not always possible. For example, we've decided that some addictive products – like heroin – are so harmful that people shouldn't be able to use them even if they want to. Other addictive products – like cigarettes – are still legal but are regulated by the state through legislation and fees. And we regulate the alcohol and gambling industries too (though not enough). It isn't difficult to imagine that we have reached a point of technological development where we need to expand the number of industries in which we heavily intervene to protect citizens against having their psychological weaknesses exploited. The big problem in this regard is that many of the superstimuli we have discussed here don't affect everyone equally. For some, life would definitely be better if food superstimuli were so heavily regulated that they were hard to come by. That would help them to maintain a healthy weight like their slim ancestors. But for many others, who are doing okay or even fine, this would limit their freedom without any benefits. It's a difficult balance. But not an impossible one to solve with some creative minds crafting legislation.

Okay, as mentioned, there is a second option to solve arms races too: change the underlying mechanics. After all, all the arms races we've encountered arise because of our inborn preferences. And they end up harming us because our preferences – what we *desire* – are no longer aligned with what benefits us in the long run. But what if we could change our preferences?

If consumers weren't asking for abnormal amounts of salt, there would be no reason to put it in products. This is where *Super Stimulated* enters the picture. Of course, not as a global solution to one of our century's greatest challenges. But as my own little input to move us all in the right direction.

Because if we are to change our preferences, the first step is to understand how we are being manipulated. Consider how smoking has gone from entirely normal to a habit for only a small part of the population. This isn't *only* because of legislation and taxes but at least as much because of information. We now know that nicotine is addictive, that smoking is terrible for your health and that tobacco companies are manipulative.

Knowledge alone might not always be enough to break a bad habit. At least not for everyone. In some cases, it can be difficult to get out of the loop, even if you know exactly how the trap works. Recall, for example, how Silicon Valley insiders desperately try to avoid the digital addiction machines they've helped to create. They might logically know what is happening but still have a hard time resisting.

So, since knowledge and willpower alone aren't always enough, it's a matter of finding tools to help. They can be the simple kind like setting your screen to greyscale and eating more potatoes, or more serious interventions like using medicine that affects the brain. I have tried to gather the best evidence I could find in this book, but there's definitely more waiting out there for you to discover.

**The United States of Superstimuli**

If you think about all the industries we have covered in this book, you might notice they have something in common: all are dominated by American companies. Food superstimuli like fast food originated in the USA and the dominant fast-food chains globally are all from there. The porn industry is headquartered in Los Angeles as are the TV, film and streaming industries. The largest computer game makers are American. All the social media companies are American. And so are all the major dating apps. Cannabis legalisation was spearheaded in America, American tobacco companies also spearheaded the rise of smoking and the opioid crisis is American too.

Now, of course, other countries make superstimuli too, my own included. But no country's exports are so thoroughly dominated by superstimuli as America's and in the global marketplace, American superstimuli often win, even if its other industries don't. It seems there's just something about God's own country that makes it particularly prone to – and good at – the predatory sides of capitalism. That's especially bad for Americans themselves, though, because being the superstimuli capital of the world means regular Americans are the number one prey and lab rats for these companies. And it shows. In obesity rates, in drug addiction, in screen time and so on.

# Epilogue: A little human in a big world

One of the brightest mathematicians ever, Srinivasa Ramanujan, grew up in South India at the end of the nineteenth century. His family was so poor that they couldn't afford school books. But as a child Ramanujan managed to teach himself mathematics by borrowing books from friends and the local library, in which he solved all the equations and then invented new ones.

The many hours of mathematics paid off when Ramanujan as a teenager was awarded a scholarship to attend university due to his exceptional abilities. Without this economic help, he would never have had the chance.

The university, however, required Ramanujan to take *other* subjects besides mathematics, such as English. But he kept failing the exams. Not because he was unintelligent or would have had difficulty keeping up with the curriculum if he had tried. Ramanujan just couldn't get himself to spend time studying anything other than maths. In the end, he had failed so many exams that he was kicked out of university, losing his scholarship and ending up back in poverty, sometimes on the verge of starving.

Ramanujan knew he had to try again and was later accepted by another university. Once again, though, he just couldn't get himself to study anything but mathematics. Eventually he had failed all the other subjects and was kicked out again.

To support himself, Ramanujan took on various odd jobs, and started tutoring university students in mathematics. The meagre pay enabled him to survive and to continue to do as much mathematics as he could on the side.

In pursuit of his dream, he sought out local maths professors to share his work. Many of them were flabbergasted when presented with the young man's theories and calculations.

Eventually, Ramanujan was discovered in the wider world, when he sent a letter to the prominent English mathematician G.H. Hardy. In the writings he described a few of his observations and included ten pages of theorems and ideas. Hardy later said that some of Ramanujan's theorems 'seemed scarcely possible to believe'. He showed them to his colleague John Littlewood, who said that the theorems were undoubtedly correct because 'no one would have had the imagination to invent them'.

After a brief correspondence, Hardy managed to bring Ramanujan to Cambridge University, where the two began working together. Unfortunately, the partnership was only brief as Ramanujan got sick after a few years and died. But in his short career he revolutionised multiple areas of mathematics, and mathematicians today continue to study his many notes to get ideas and make new discoveries.[1]

Ramanujan's relationship to mathematics resembles the relationship other people have to computer games, alcohol or gambling. Even when his entire existence depended on it, he had difficulty forcing himself to deal with other things. With a little effort Ramanujan could have kept his university scholarship and much more easily made a career for himself in mathematics, as he had dreamed. But even though he could study mathematics for days on end, he was simply unable to muster the effort for any other subject.

This might be an extreme example, but there are echoes of Ramanujan's story among many successful people. There is often an element of obsession in high achievers. An example from another discipline could be the Hungarian Polgar sisters, who are the best female chess players the world has ever seen. Their father, Laszlo Polgar, has described how the sisters were obsessed with chess as children, and entirely uninterested in doing anything else. One night he found one of them, Sofia, in the bathroom with the chess board. 'Sofia, leave the pieces alone!' he insisted. 'Daddy,' she answered, 'they won't leave *me* alone!'[2]

You might also recognise this same behaviour from your own life. At least, everyone I know who is great at something shares a strong natural interest in whatever they do. It isn't always there from the start. Some need to get a feel for their field before their obsession kickstarts. But they are all obsessed.

That we see this behaviour isn't so strange. Of course, you need talent to go far. But success takes a lot of work and it is easier to work a lot if you don't have to force yourself but can spend the hours effortlessly absorbed.

*

We don't know exactly why some activities feel more strenuous than others. In some cases, it makes sense. When you run, you burn more calories than you do lying in bed. Insofar as your body is built to conserve energy, of course lying in bed feels nicer than running.

But why is it more strenuous to solve mathematics equations than to play a computer game?

It can't be because of energy consumption. Your brain does burn a lot of calories – close to 20 per cent of the daily energy consumption – but this number doesn't increase much when you are thinking deeply or concentrating.[3] So, when it comes to mental tasks, the feeling of strain is a bit of a mystery. It doesn't take more energy to sit still and solve an equation than to watch a movie.

A group of American psychologists have a theory that might help us explain the discrepancy. They hypothesise that the feeling of strain comes from the brain trying to get us to do the most rewarding activity at any given time. Imagine that your brain constantly evaluates all the potential things you could be doing at the moment and weighs them according to how rewarding it expects them to be. If the reward from whatever you're currently doing is lower than the reward from something you could be doing instead, you need to exert effort to keep going. But if the reward from your current activity is greater, you have no problem continuing.

This theory would mean superstimuli are a problem for anyone with goals and aspirations. If you become accustomed to enormously stimulating TikTok feeds, computer games and Netflix series, these are the activities that your brain is comparing other activities against. Those activities will then take more willpower because they are not as instantly rewarding as the artificially designed superstimuli. Consider how many modern people set self-improvement goals like starting to paint, learning to play an instrument or reading more books. These are all activities that people used to do *for fun*. If you were living 200 years ago and needed a way to spend your free time, these activities would be among the most rewarding choices. But today they fall into the category of activities that require willpower for many because they have to compete with modern superstimuli. Similarly, some young people today complain of being unable to watch a whole film from start to finish. It might have been the previous generation's superstimulus, but in competition with TikTok-like social media feeds, it loses out and ends up as something requiring effort to do.

We see an extreme example of this phenomenon among people with substance use disorders. As their addictions grow, they lose interest in their old activities and stop caring about their friends and family. However, people suffering from drug addiction will exert almost superhuman efforts to obtain their substance. This makes sense if their brain sees taking the substance as the absolutely most rewarding thing the addicted person could do. Other activities become immensely strenuous and impossible to do, but the hunt for a high *doesn't*, even though it would be enormously hard for a non-addicted person, sometimes involving committing violence, theft or putting your life on the line.

This 'comparison' theory of willpower also aligns with studies in which scientists investigate people's willingness to exert themselves. Usually, these studies show that people prefer to complete tasks that require minimal effort. Doing mathematical problems isn't a popular choice when people are presented with multiple options.[4] But when scientists present people with the choice between doing maths problems and doing nothing, many choose the maths problems over staring into space.[5] In other words, whether maths problems are strenuous or not depends on what the alternative is.

★

It's hard not to wonder how many people are sitting out there, neglecting their talents and dreams because superstimuli have made everything but artificial overstimulation difficult to bear. How many present-day Ramanujan's are fanatical *World of Warcraft* players instead of maths nerds?

And more frightening: how bad will this get in the future?

You see, unfortunately, these developments aren't over yet. Once upon a time, the most stimulating computer game was Pong, a game in which you move two lines up and down to shoot a ball back and forth across the screen. Today we have games that take place in enormously detailed fictional worlds, which give the player nearly unlimited options. In other domains, we've gone from pictures of scantily clad men and women to internet streaming with endless hardcore pornography. From cannabis with a little THC to crystals containing almost pure THC. And from vanilla ice cream to the infamous vanilla ice cream with caramel sauce, chocolate chips, marshmallows and cookie dough.

In all these areas, there are undoubtedly new escalations out there waiting to be discovered. Particularly in the digital world, developments are proceeding full steam ahead. In the realm of bits, superstimuli makers have it especially easy because doing experiments is quick, cheap and simple to set up. Food manufacturers need to physically produce all the different versions of their product and then test them in experiments in the real world. But as we discussed, digital companies can run thousands of small experiments on you every day because their products are easily adjustable computer codes.

At the same time, digital companies are increasingly assisted by artificial intelligence, so that humans don't even need to perform or be involved in the experiments at all. Soon, the very generation of the content might be automated as well, so that artificial intelligence doesn't only study the user to curate the feed but generates all the videos, texts, people and so on in a way that is tailored for them. When this happens, I expect a portion of society will pretty much fully withdraw to the digital world. But while that is sad and tragic, it doesn't mean the future is hopeless for all.

First, because producers of superstimuli aren't the only ones taking advantage of new and improved technology. Our *defences* also improve and there is no reason they couldn't get an upper hand. That's a way of stopping an arms race I left out before. You can also stop an arms race by winning it. Think back to the chapter on weight loss drugs. Once upon a time, willpower and behavioural tricks were our only weapons against food superstimuli. This wasn't enough for most people and the fight was continuously escalating as the food industry discovered novel tricks. But now medicine has put the consumer back on top. The food industry might be scrambling to regain the upper hand but it is not a given that this is even possible and the drugs will likely also improve from here.

Indeed, scientists are now studying different GLP-1 and related drugs to find out whether they can help people shake *other* bad habits as well. In studies on rodents and non-human primates, some of these medications reduce the desire for both drugs and alcohol.[6] Some early studies in humans point in the same direction, though there is undoubtedly more to learn.[7] It's probably too starry-eyed to believe that a single category of medication can solve all our superstimuli problems, but we can hope that it will be joined by more sources of progress in the future.

In this discussion, it is also interesting to note that what technology is actually doing, is not leading everyone into despair but rather accelerating and accentuating existing differences between people. The space between the top and the bottom grows. Take our good friends the Hadza, who largely live without modern technology. They are slimmer than the average Westerner, more muscular and healthier overall in many ways. But it is important to emphasise the word *average*. There are plenty of people in the West who are stronger than the strongest Hadza, who can run faster and longer than the Hadza, who carry more muscle mass than the Hadza, and who are just healthier on every single parameter: metabolism, general health, lifespan and so on.

This is because modern technology lets us optimise our physiques in ways that traditional populations could only dream of. So, it makes sense to view technological developments as a foot on the accelerator; it will let you accelerate the speed, no matter what direction, you're heading.

If you follow your instincts, there is an army of food manufacturers ready to exploit your brain's reward system. Their products can make you overweight and weak, to such a degree that you wouldn't have survived just a few hundred years ago. On the other hand, our knowledge of biology and medicine has also become so great that we can physically achieve things that would have been entirely impossible for our forefathers.

Similarly, the rise of the digital world can be used to watch random videos for hours on end but it can also be used as it was originally intended, to access all the knowledge of the world in a second. Even from a desk in a sleepy village, thanks to modern technology, not a single discipline is out of reach. So, while there are people today who are more ignorant, isolated and inactive than ever before, there are also people who know and are able to achieve more than our ancestors could ever dream of. The question of whether technological change is good or bad depends on your perspective.

Yes, even superstimuli aren't necessarily *bad*.

You can imagine situations in which we could use all that we've learned about superstimuli to create something good. We could design teaching methods or technologies that could be as captivating for students as TikTok.

Or what if we could use our knowledge about our appetite to design processed foods that are *healthier* than natural foods and more nourishing?

What determines whether superstimuli are good or bad is what they make us do. If they lead to ruin, we must fight against them. But if we are smart, we learn to use them for our own good.

While reading, maybe you've noticed that's what has been helping you reach the end of this book.

# Acknowledgements

A book like this has a sole author on the cover but behind the scenes there are many brilliant people involved that each deserve a literary applause. Anna Baty and Lucy Buxton had to endure my perfectionism but still managed to lift the book to the next level. Also thank you to Sherilyn Hellberg for translating this book from Danish into English, to Tamsin Shelton for copyediting the end product, to Paul Sebes and colleagues at my agency Sebes & Bisseling for handling the deal and to the many editors, translators and proof-readers who are working on the other translations of the book.

*Super Stimulated* started out in Danish and my Danish publishers at Grønningen 1, Louise Vind and Marianne Kiertzner, deserve a special thanks. When I wrote my first book, *Jellyfish Age Backwards*, every single major publisher in the country rejected it. Some of them rejected it twice because I was a little stubborn and didn't take no for an answer. In the end, I pretty much ran out of options and was on the verge of having to ask my parents to borrow money to self-publish. But fortunately, I met Louise and Marianne who believed in the project right away and took on the book. Now, it has sold hundreds of thousands of copies worldwide, been nominated for Science Book of the Year by the Royal Society and been translated into 30 different languages. I still keep the emails from other publishers explaining in thorough detail why there was no market for the book and why it couldn't possibly succeed, defined in Denmark as selling a few thousand copies in total.

I also want to thank my scientific colleagues at Trygfonden's Centre for Physical Activity Research at Rigshospitalet in Copenhagen. This especially goes for my PhD supervisors Bente Klarlund Pedersen and Helga Ellingsgaard who allow me more freedom to pursue my writing and speaking than any PhD student could reasonably expect.

Additionally, a special mention to my podcast co-host Frederik Dirks Gottlieb and everyone else at Danmarks Radio involved in *Jagten på det evige liv*. And to all my colleagues at Original Talks, Athenas and A-Speakers who help arrange my talks so that I have to leave the laboratory and writing desk occasionally.

I also want to express my gratitude to all my friends, near and far, for their support and for making my life richer. An extra touch of appreciation to Eske Gerup, Joachim Schelde, Lars Horsbøl, Mikkel Duif and Sven Larsen-Ledet for all the great adventures, at home and abroad.

Finally, I want to send a special thanks to my family for whom my love is greater than what can ever be described in words and to whom I dedicate this book and all others.

# Notes

## Prologue: A small bird with a large egg

1. Tinbergen N. *The Study of Instinct*. Clarendon Press. 1951, 1989.
2. Moreno J., Morales J., Lobato E., Merino S., Tomás G., Martínez-de la Puente J. 'More colourful eggs induce a higher relative paternal investment in the pied flycatcher Ficedula hypoleuca: a cross-fostering experiment'. *Journal of Avian Biology*. 2006;37(6):555–60.
3. Moreno J., Lobato E., Merino S., Martínez-de la Puente J. 'Blue-Green Eggs in Pied Flycatchers: An Experimental Demonstration that a Supernormal Stimulus Elicits Improved Nestling Condition'. *Ethology*. 2008;114(11):1078–83.

## Part 1: FOOD SUPERSTIMULI

### 1. The weight of modernity

1. Floud R. 'Height, Weight and Body Mass of the British Population Since 1820'. NBER Working Paper Series on Historical Factors in Long Run Growth. Historical Paper 108. October 1998. https://www.nber.org/system/files/working_papers/h0108/h0108.pdf.
2. Health Survey for England 2022, NHS Digital. https://digital.nhs.uk/data-and-information/publications/statistical/health-survey-for-england/2022-part-2/health-survey-for-england-hse-2022-part-2-data-tables.
3. 2024 State of the UK Fitness Industry Report (Leisure DB).

4. Madsen B.F. 'Forskere afslører: BMI-skalaen har blindt punkt hos stor aldersgruppe'. *Illustreret Videnskab.* 13.06.24. https://illvid.dk/sundhed/forskere-afsloerer-bmi-skalaen-har-blindt-punkt-hos-stor-aldersgruppe.
5. Overweight and obese adults, 2018. Statistics Canada. https://www150.statcan.gc.ca/n1/pub/82-625-x/2019001/article/00005-eng.htm.
6. Overweight and obesity. Australian Institute of Health and Welfare. https://www.aihw.gov.au/reports/overweight-obesity/overweight-and-obesity/contents/summary#how_common_overweight.
7. Annual Update of Key Results 2023/24: New Zealand Health Survey. https://www.health.govt.nz/publications/annual-update-of-key-results-202324-new-zealand-health-survey.
8. Baker C. Research Briefing: Obesity statistics. House of Commons Library. 12.1.23. https://commonslibrary.parliament.uk/research-briefings/sn03336/.
9. Prevalence of obesity among adults. World Health Organization. https://www.who.int/data/gho/data/indicators/indicator-details/GHO/prevalence-of-obesity-among-adults-bmi--30-(crude-estimate)-(-).
10. Ibid.
11. Turé R., Damasceno A., Djicó M., Lunet N. 'Prevalence of Underweight, Overweight and Obesity among Adults in Urban Bissau, Western Africa'. *Nutrients.* 2021 Nov 23;13(12):4199.
12. Pengpid S. and Peltzer K. 'The prevalence and associated factors of underweight and overweight/obesity among adults in Kenya: evidence from a national cross-sectional community survey'. *Pan African Medical Journal.* 2020 Aug 25;36:338.
13. Letamo G. 'Dual burden of underweight and overweight/obesity among adults in Botswana: prevalence, trends and sociodemographic correlates: a cross-sectional survey'. *British Medical Journal.* 2020;10(7).
14. Prevalence of obesity among adults. World Health Organization. https://www.who.int/data/gho/data/indicators/indicator-details/GHO/prevalence-of-obesity-among-adults-bmi--30-(crude-estimate)-(-).
15. Martin C.B., Herrick K.A., Sarafrazi N., Ogden C.L. 'Attempts to Lose Weight Among Adults in the United States, 2013–2016'. *NCHS Data Brief.* 2018(313):1–8.

16. Hall K.D. and Kahan S. 'Maintenance of lost weight and long-term management of obesity'. *Medical Clinics of North America.* 2018 Jan;102(1):183–197.
17. Althoff T., Sosič R., Hicks J.L., King A.C., Delp S.L., Leskovec J., 'Supplementary Information: Large-scale physical activity data reveal worldwide activity inequality'. *Nature.* 2017 Jul 10;547(7663):336–339.
18. Pontzer H., Raichlen D.A., Wood B.M., Mabulla A.Z., Racette S.B., Marlowe F.W. 'Hunter-gatherer energetics and human obesity'. *PLoS One.* 2012;7(7):e40503.
19. Pontzer H., Wood B.M., Raichlen D.A. 'Hunter-gatherers as models in public health'. *Obesity Reviews.* 2018;19(S1):24–35.
20. Urlacher S.S., Snodgrass J.J., Dugas L.R., Sugiyama L.S., Liebert M.A., Joyce C.J., et al. 'Constraint and trade-offs regulate energy expenditure during childhood'. *Science Advances.* 5(12):eaax1065.
21. Pontzer H., Raichlen D.A., et al. 'Primate energy expenditure and life history'. *Biological Sciences.* 13 January 2014. 111(4):1433-1437.
22. Pontzer, H. *Burn.* Avery Publishing. 2021.
23. Pontzer H., Raichlen D.A., Gordon A.D., Schroepfer-Walker K.K., Hare B., O'Neill M.C., et al. 'Primate energy expenditure and life history'. *Proc Natl Acad Sci USA.* 2014;111(4):1433–7.
24. Careau V., Halsey L.G., Pontzer H., Ainslie P.N., Andersen L.F., Anderson L.J., et al. 'Energy compensation and adiposity in humans'. *Curr Biol.* 2021;31(20):4659–66.e2.
25. Pontzer, H. *Burn.* Avery Publishing. 2021. pp.243–244.
26. Bribiescas R.G. 'Testosterone levels among Aché hunter-gatherer men'. *Human Nature.* 1996;7(2):163–88.
27. Melanson E.L., Keadle S.K., Donnelly J.E., Braun B., King N.A. 'Resistance to exercise-induced weight loss: compensatory behavioral adaptations'. *Med Sci Sports Exerc.* 2013;45(8):1600–9.
28. Donnelly J.E., Honas J.J., Smith B.K., Mayo M.S., Gibson C.A., Sullivan D.K., et al. 'Aerobic exercise alone results in clinically significant weight loss for men and women: midwest exercise trial 2'. *Obesity* (Silver Spring). 2013;21(3):E219–28.
29. Westerterp K.R. and Speakman J.R. 'Physical activity energy expenditure has not declined since the 1980s and matches energy expenditures of wild mammals'. *International Journal of Obesity.* 2008;32(8):1256–63.

## 2. Hijacking hunger

1. Stunkard A.J., Sørensen T.I.A., Hanis C., Teasdale T.W., Chakraborty R., Schull W.J., et al. 'An Adoption Study of Human Obesity'. *New England Journal of Medicine*. 1986;314(4):193–8.
2. Stunkard A.J., Harris J.R., Pedersen N.L., McClearn G.E. 'The Body-Mass Index of Twins Who Have Been Reared Apart'. *New England Journal of Medicine*. 1990;322(21):1483–7.
3. Yengo L., Sidorenko J., Kemper K.E., Zheng Z., Wood A.R., Weedon M.N., et al. 'Meta-analysis of genome-wide association studies for height and body mass index in ~700000 individuals of European ancestry'. *Hum Mol Genet*. 2018;27(20):3641–9.
4. FAO. Value of Agricultural Production. FAO.org. [updated 2023-03-21].
5. Swinburn B., Sacks G., Ravussin E. 'Increased food energy supply is more than sufficient to explain the US epidemic of obesity'. *The American Journal of Clinical Nutrition*. 2009;90(6):1453–6.
6. Moss, M. *Salt Sugar Fat: How the Food Giants Hooked Us*. Random House. 2014. pp.150–151.
7. Hall K.D., Ayuketah A., Brychta R., Cai H., Cassimatis T., Chen KY., et al. 'Ultra-Processed Diets Cause Excess Calorie Intake and Weight Gain: An Inpatient Randomized Controlled Trial of Ad Libitum Food Intake'. *Cell Metab*. 2019;30(1):67-77.e3.
8. Hamano S., Sawada M., Aihara M., Sakurai Y., Sekine R., Usami S., Kubota N., Yamauchi T. 'Ultra-processed foods cause weight gain and increased energy intake associated with reduced chewing frequency: A randomized, open-label, crossover study'. *Diabetes, Obesity and Metabolism*. 2024;26(11):5431–43.
9. Sampey B.P., Vanhoose A.M., Winfield H.M., Freemerman A.J., Muehlbauer M.J., Fueger P.T., Newgard C.B., Makowski L. 'Cafeteria diet is a robust model of human metabolic syndrome with liver and adipose inflammation: comparison to high-fat diet'. *Obesity* (Silver Spring). 2011;19(6):1109–17.

## 3. Tales of the white gold

1. Allain F., Minogianis E., Roberts D.C.S., Samaha A. 'How fast and how often: The pharmacokinetics of drug use are decisive in addiction'. *Neuroscience and Biobehavioral Reviews.* 2015 Sept:166–179.
2. Headey D.D. and Alderman H.H. 'The Relative Caloric Prices of Healthy and Unhealthy Foods Differ Systematically across Income Levels and Continents'. *J Nutr.* 2019;149(11):2020–33.
3. Public Health England. *National Diet and Nutrition Survey. Rolling programme Years 9 to 11 (2016/2017 to 2018/2019).* December 2020. https://assets.publishing.service.gov.uk/media/5fd23324e90e07662b09d91a/NDNS_UK_Y9-11_report.pdf.
4. Langlois K., Garriguet D. et al., 'Change in total sugars consumption among Canadian children and adults'. *Health Rep.* 2019 Jan 16;30(1):10–19.
5. Pesta D.H. and Samuel V.T., 'A high-protein diet for reducing body fat: mechanisms and possible caveats'. *Nutr Metab (Lond).* 2014 Nov 19;11:53.
6. Yu Y., Fu J., Xu Y., Zhang J., Ren F., Zhao H., et al. 'Genome re-sequencing reveals the evolutionary history of peach fruit edibility'. *Nature Communications.* 2018;9(1):5404.
7. Zheng Y., Crawford G.W., Chen X. 'Archaeological evidence for peach (Prunus persica) cultivation and domestication in China'. *PLoS One.* 2014;9(9):e106595.
8. Quilot B., Kervella J., Génard M. 'Shape, mass and dry matter content of peaches of varieties with different domestication levels'. *Scientia Horticulturae.* 2004;99(3):387–93.
9. Pérez-Escobar O., et al. 'Genome Sequencing of up to 6,000-Year-Old Citrullus Seeds Reveals Use of a Bitter-Fleshed Species Prior to Watermelon Domestication'. *Molecular Biology and Evolution.* 2022;39(8).
10. Malik V.S. and Hu F.B. 'The role of sugar-sweetened beverages in the global epidemics of obesity and chronic diseases'. *Nat Rev Endocrinol.* 2022;18:205–18.
11. DellaValle D.M., Roe L.S., Rolls B.J. 'Does the consumption of caloric and non-caloric beverages with a meal affect energy intake?' *Appetite.* 2005;44(2):187–93.

12. Aeberli I., Gerber P.A., Hochuli M., Kohler S., Haile S.R., Gouni-Berthold I., et al. 'Low to moderate sugar-sweetened beverage consumption impairs glucose and lipid metabolism and promotes inflammation in healthy young men: a randomized controlled trial'. *The American Journal of Clinical Nutrition*. 2011;94(2):479–85.
13. Buchanan K.L., Rupprecht L.E., Kaelberer M.M., Sahasrabudhe A., Klein M.E., Villalobos J.A., et al. 'The preference for sugar over sweetener depends on a gut sensor cell'. *Nature Neuroscience*. 2022;25(2):191–200.
14. Tan H., Sisti A., Jin H. et al., 'The gut-brain axis mediates sugar preference'. *Nature*. 2020;580:511–516.
15. Jones S., Sullivan H. 'Subway bread is not bread, Irish court rules'. *The Guardian* [online]. 1.10.20.
16. Saccharin, Britannica.com. https://www.britannica.com/science/saccharin#ref165773.
17. Roth K. and Lück E. 'The Saccharin Saga – Part 7'. *Chemistry Views*. 3.5.16.
18. Higgins K.A. and Mattes R.D. 'A randomized controlled trial contrasting the effects of 4 low-calorie sweeteners and sucrose on body weight in adults with overweight or obesity'. *Am J Clin Nutr*. 2019;109(5):1288–301.
19. McGlynn N.D., Khan T.A., Wang L., Zhang R., Chiavaroli L., Au-Yeung F., et al. 'Association of Low- and No-Calorie Sweetened Beverages as a Replacement for Sugar-Sweetened Beverages With Body Weight and Cardiometabolic Risk: A Systematic Review and Meta-analysis'. *JAMA Network Open*. 2022;5(3):e222092-e.
20. Ebbeling C.B., Feldman H.A., Steltz S.K., Quinn N.L., Robinson L.M., Ludwig D.S. 'Effects of Sugar-Sweetened, Artificially Sweetened, and Unsweetened Beverages on Cardiometabolic Risk Factors, Body Composition, and Sweet Taste Preference: A Randomized Controlled Trial'. *Journal of the American Heart Association*. 2020;9(15):e015668.
21. World Health Organization. 'Aspartame hazard and risk assessment results released'. 14.7.23. https://www.who.int/news/item/14-07-2023-aspartame-hazard-and-risk-assessment-results-released.
22. Magnuson B.A., Burdock G.A., Doull J., Kroes R.M., Marsh G.M., et al. 'Aspartame: A Safety Evaluation Based on Current Use Levels, Regulations, and Toxicological and Epidemiological Studies'. *Critical Reviews in Toxicology*. 2007;37(8):629–727.

23. Magnuson B.A., Carakostas M.C., Moore N.H., Poulos S.P., Renwick A.G. 'Biological fate of low-calorie sweeteners'. *Nutrition Reviews*. 2016;74(11):670–89.

## 4. Rabbit starvation

1. Force USAA. *Jungle, Desert, and Arctic Emergencies Booklet*. Laupus Library History Collections: Flight Control Command Safety Education Division of the United States Army Air Forces; 1947.
2. Bella S.D., Branciari R., et al. 'Does hunted wild boar meat meet modern consumer nutritional expectations?'. *Ital J Food Saf.* 2024 Feb 22;13(1):11608.
3. Bison vs Beef: What's the Difference? healthline. https://www.healthline.com/nutrition/bison-vs-beef.
4. Jaime-Lara R.B., Brooks B.E.,Vizioli C., et al. 'A systematic review of the biological mediators of fat taste and smell'. *Physiological Reviews*. 2023;103(1):855–918.
5. Mengtong, L., Tan, H., Zhengyuan L. et al. 'Gut-brain circuits for fat preference'. *Nature*. 2022;610:722–730.
6. DiFeliceantonio A.G., Coppin G., Rigoux L., Edwin Thanarajah S., Dagher A., Tittgemeyer M., et al. 'Supra-Additive Effects of Combining Fat and Carbohydrate on Food Reward'. *Cell Metab*. 2018;28(1):33–44.e3.
7. Roberts S.B., Das S.K., Suen V.M.M., Pihlajamäki J., Kuriyan R., Steiner-Asiedu M., et al. 'Measured energy content of frequently purchased restaurant meals: multi-country cross sectional study'. *BMJ*. 2018;363:k4864.
8. Drewnowski A., Schwartz M. 'Invisible fats: Sensory assessment of sugar/fat mixtures'. *Appetite*. 1990;14(3):203–17.
9. Ibid.
10. Kim S.Y., Yi D.Y. 'Components of human breast milk: from macronutrient to microbiome and microRNA'. *Clin Exp Pediatr*. 2020;63(8):301–9.
11. Pontzer H., Yamada Y., Sagayama H., Ainslie P.N., et al. 'Daily energy expenditure through the human life course.' *Science*. 2021 August 13;373(6556):808–12.

## 5. What skydiving and salt have in common

1. Campbell N.R.C., Correa-Rotter R., et al. 'Proposed Nomenclature for Salt Intake and for Reductions in Dietary Salt'. *J Clin Hypertens (Greenwich)*. 2014 Nov 21;17(4):247–251.
2. Land M., Neal B.C., et al. 'Salt consumption by Australian adults: a systematic review and meta-analysis'. *Med J Aust*. 2018 Feb 5;208(2):75–81.
3. Public Health England. *National Diet and Nutrition Survey: Assessment of salt intake from urinary sodium in adults (aged 19 to 64 years) in England, 2018 to 2019*. March 2020.
4. Bertino M., Beauchamp G.K., Engelman K. 'Long-term reduction in dietary sodium alters the taste of salt'. *The American Journal of Clinical Nutrition*. 1982;36(6):1134–44.
5. Wise P.M., Nattress L., Flammer L.J., Beauchamp G.K. 'Reduced dietary intake of simple sugars alters perceived sweet taste intensity but not perceived pleasantness'. *The American Journal of Clinical Nutrition*. 2016;103(1):50–60.
6. Lee R. *The !Kung-San: Men, Women and Work in a Foraging Society*. Cambridge University Press. 1980.
7. Roe L.S., Meengs J.S., Birch L.L., Rolls B.J. 'Serving a variety of vegetables and fruit as a snack increased intake in preschool children'. *The American Journal of Clinical Nutrition*. 2013;98(3):693–9.

## 6. Health advice from the Amazon

1. Kaplan H., Thompson R.C., Trumble B.C., Wann L.S., Allam A.H., Beheim B., et al. 'Coronary atherosclerosis in indigenous South American Tsimane: a cross-sectional cohort study'. *The Lancet*. 2017;389(10080):1730–9.
2. Roth G.A., Abate D., Abate K.H., Abay S.M., Abbafati C., Abbasi N., et al. 'Global, regional, and national age-sex-specific mortality for 282 causes of death in 195 countries and territories, 1980–2017: a systematic analysis for the Global Burden of Disease Study 2017'. *The Lancet*. 2018;392(10159):1736–88.
3. Libby P., Buring J.E., et al. 'Atherosclerosis'. *Nature Reviews Disease Primers*. 2019 Aug;5:56.

4. Global Burden of Disease 2021, https://www.healthdata.org/sites/default/files/2024-05/GBD_2021_Booklet_FINAL_2024.05.16.pdf.
5. Sharif K., Amital H., Shoenfeld Y. 'The role of dietary sodium in autoimmune diseases: The salty truth'. *Autoimmunity Reviews.* 2018;17(11):1069–73.
6. Filippini T., Malavolti M., Whelton P.K., Vinceti M. 'Sodium Intake and Risk of Hypertension: A Systematic Review and Dose-Response Meta-analysis of Observational Cohort Studies'. *Curr Hypertens Rep.* 2022;24(5):133–44.
7. Campbell N.R., Correa-Rotter R., Cappuccio F.P., Webster J., Lackland D.T., Neal B., et al. 'Proposed nomenclature for salt intake and for reductions in dietary salt'. *J Clin Hypertens* (Greenwich). 2015;17(4):247–51.
8. Denton D., Weisinger R., Mundy N.I., Wickings E.J., Dixson A., Moisson P., et al. 'The effect of increased salt intake on blood pressure of chimpanzees'. *Nature Medicine.* 1995;1(10):1009–16.
9. Hunter R.W., Dhaun N., Bailey M.A. 'The impact of excessive salt intake on human health'. *Nature Reviews Nephrology.* 2022;18(5):321–35.
10. Lemogoum D., Ngatchou W., Janssen C., Leeman M., Van Bortel L., Boutouyrie P., et al. 'Effects of Hunter-Gatherer Subsistence Mode on Arterial Distensibility in Cameroonian Pygmies'. *Hypertension.* 2012;60(1):123–8.
11. Ference, B.A., Ginsberg, H.N., et al. 'Low-density lipoproteins cause atherosclerotic cardiovascular disease. 1. Evidence from genetic, epidemiologic, and clinical studies. A consensus statement from the European Atherosclerosis Society Consensus Panel'. *Eur Heart J.* 2017 Aug 21;38(32):2459–2472.
12. Vasunilashorn, S., Crimmins, E.M., Kim, J.K., et al. 'Blood lipids, infection, and inflammatory markers in the Tsimane of Bolivia'. *American Journal of Human Biology.* 2020 Aug;22:731–740.
13. O'Keefe J.H., Cordain L., Harris W.H., Moe R.M., Vogel R. 'Optimal low-density lipoprotein is 50 to 70 mg/dl: Lower is better and physiologically normal'. *Journal of the American College of Cardiology.* 2004;43(11):2142–6.
14. Kraft T.S., Stieglitz J., Trumble B.C., Martin M., Kaplan H., Gurven M. 'Nutrition transition in 2 lowland Bolivian subsistence populations'. *Am J Clin Nutr.* 2018;108(6):1183–95.

15. Mensink R.P., World Health Organization. 'Effects of saturated fatty acids on serum lipids and lipoproteins: a systematic review and regression analysis'. World Health Organization. 2016.
16. Pedersen A.N., Christensen, T., Matthiessen J., et al. 'Danskernes kostvaner 2011–2013'. DTU Fødevareinstituttet. 2015.
17. King D.E., Mainous A.G., Lambourne C.A. 'Trends in Dietary Fiber Intake in the United States, 1999–2008'. *Journal of the Academy of Nutrition and Dietetics*. 2012;112(5):642–8.
18. Raichlen D.A., Pontzer H., Harris J.A., Mabulla A.Z., Marlowe F.W., Snodgrass J., et al. 'Physical activity patterns and biomarkers of cardiovascular disease risk in hunter-gatherers'. *Am J Hum Biol*. 2017;29(2).
19. Reynolds A., Mann J., Cummings J., Winter N., Mete E., Te Morenga L. 'Carbohydrate quality and human health: a series of systematic reviews and meta-analyses'. *The Lancet*. 2019;393(10170):434–45.
20. Jovanovski E., Yashpal S., Komishon A., Zurbau A., Blanco Mejia S., Ho H.V.T., et al. 'Effect of psyllium Plantago ovata fiber on LDL cholesterol and alternative lipid targets, non-HDL cholesterol and apolipoprotein B: a systematic review and meta-analysis of randomized controlled trials'. *The American Journal of Clinical Nutrition*. 2018;108(5):922–32.
21. Cicero A.F.G., Fogacci F., Veronesi M., Strocchi E., Grandi E., Rizzoli E., et al. 'A Randomized Placebo-Controlled Clinical Trial to Evaluate the Medium-Term Effects of Oat Fibers on Human Health: The Beta-Glucan Effects on Lipid Profile, Glycemia and inTestinal Health (BELT) Study'. *Nutrients* [online]. 2020;12(3).
22. Kern F. 'Normal Plasma Cholesterol in an 88-Year-Old Man Who Eats 25 Eggs a Day – Mechanisms of Adaptation'. *The New England Journal of Medicine*. 1991;324:896–899.
23. Soliman G.A. 'Dietary Cholesterol and the Lack of Evidence in Cardiovascular Disease'. *Nutrients*. 2018 Jun 16;10(6):780.
24. Gurven M. and Hillard K. 'Longevity Among Hunter-Gatherers: A Cross-Cultural Examination'. Gurven Lab. https://www.gurven.anth.ucsb.edu/sites/secure.lsit.ucsb.edu.anth.d7_gurven/files/sitefiles/papers/GurvenKaplan2007pdr.pdf.
25. Gurven M. and Kaplan H. 'Longevity Among Hunter-Gatherers: A Cross-Cultural Examination'. *Population and Development Review*. 2007;33:321–365.

26. Arranz-Otaegui A., Gonzalez Carretero L., Ramsey M.N., Fuller D.Q., Richter T. 'Archaeobotanical evidence reveals the origins of bread 14,400 years ago in northeastern Jordan'. *Proceedings of the National Academy of Sciences*. 2018;115(31):7925–30.
27. Piperno D.R., Weiss E., Holst I., Nadel D. 'Processing of wild cereal grains in the Upper Palaeolithic revealed by starch grain analysis'. *Nature*. 2004;430(7000):670–3.
28. Henry A.G., Brooks A.S., Piperno D.R. 'Microfossils in calculus demonstrate consumption of plants and cooked foods in Neanderthal diets (Shanidar III, Iraq; Spy I and II, Belgium)'. *Proceedings of the National Academy of Sciences*. 2011;108(2): 486–91.

## 7. The spud diet

1. Interview with the author.
2. Holt S.H.A., Brand Miller J.C., Petocz P., Farmakalidis E. 'A Satiety Index of Common Foods'. *European Journal of Clinical Nutrition*. September 1995:675-90.
3. Ku S.K., Sung S.H., Choung J.J., Choi J.S., Shin Y.K., Kim J.W. 'Anti-obesity and anti-diabetic effects of a standardized potato extract in ob/ob mice'. *Exp Ther Med*. 2016;12(1):354–64.
4. Hill A.J., Peikin S.R., Ryan C.A., Blundell J.E. 'Oral administration of proteinase inhibitor II from potatoes reduces energy intake in man'. *Physiology & Behavior*. 1990;48(2):241–6.
5. Klos B., Cook J., Crepaz L., Weiland A., Zipfel S., Mack I. 'Impact of energy density on energy intake in children and adults: a systematic review and meta-analysis of randomized controlled trials'. *European Journal of Nutrition*. 2023;62(3):1059–76.
6. Sobiecki J.G., Appleby P.N., Bradbury K.E., Key T.J. 'High compliance with dietary recommendations in a cohort of meat eaters, fish eaters, vegetarians, and vegans: results from the European Prospective Investigation into Cancer and Nutrition–Oxford study'. *Nutrition Research*. 2016;36(5):464–77.
7. Tonstad S., Butler T., Yan R., Fraser G.E. 'Type of vegetarian diet, body weight, and prevalence of type 2 diabetes'. *Diabetes Care*. 2009;32(5):791–6.

8. Johnston B.C., Kanters S., Bandayrel K., Wu P., Naji F., Siemieniuk R.A., et al. 'Comparison of Weight Loss Among Named Diet Programs in Overweight and Obese Adults: A Meta-analysis'. *JAMA*. 2014;312(9):923–33.
9. Pontzer, H. *Burn*. Avery Publishing. 2021.
10. Voss J.D., Masuoka P., Webber B.J., Scher A.I., Atkinson R.L. 'Association of elevation, urbanization and ambient temperature with obesity prevalence in the United States'. *International Journal of Obesity*. 2013;37(10):1407–12.
11. Díaz-Gutiérrez J., Martínez-González M., Pons Izquierdo J.J., González-Muniesa P., Martínez J.A., Bes-Rastrollo M. 'Living at Higher Altitude and Incidence of Overweight/Obesity: Prospective Analysis of the SUN Cohort'. *PLoS One*. 2016;11(11):e0164483.
12. Voss J.D., Allison D.B., Webber B.J., Otto J.L., Clark L.L. 'Lower obesity rate during residence at high altitude among a military population with frequent migration: a quasi experimental model for investigating spatial causation'. *PLoS One*. 2014;9(4):e93493.
13. Burdack-Freitag A., Bullinger D., Mayer F., Breuer K. 'Odor and taste perception at normal and low atmospheric pressure in a simulated aircraft cabin'. *Journal für Verbraucherschutz und Lebensmittelsicherheit*. 2010;6:95–109.

## 8. From dynamite to diet pills

1. Perkins R.G. 'A Study of the Munitions Intoxications in France'. Public Health Report 1919;34(43):2335–74.
2. Tainter M.L., Cutting W.C., Stockton A.B. 'Use of Dinitrophenol in Nutritional Disorders: A Critical Survey of Clinical Results'. *Am J Public Health Nations Health*. 1934;24(10):1045–53.
3. Horner W.D. 'A Study of Dinitrophenol and Its Relation to Cataract Formation'. *Trans Am Ophthalmol Soc.* 1941;39:405–37.
4. Pi-Sunyer X., Aronne L.J., Heshmati H.M. et al. 'Effect of Rimonabant, a Cannabinoid-1 Receptor Blocker, on Weight and Cardiometabolic Risk Factors in Overweight or Obese Patients'. *JAMA*. 2006;295(7):761–775.
5. Saul S. 'F.D.A. Panel Reject Drug for Obesity'. 14.06.07: https://www.nytimes.com/2007/06/14/business/14drugs.html.

6. Hughes J. 'Anti-obesity drug use suspended'. BBC News. 23.10.08: http://news.bbc.co.uk/2/hi/health/7687311.stm.
7. Allchurch M.H. 'The European Medicines Agency recommends suspension of the marketing authorisation of Acomplia'. 23.10.08: https://www.ema.europa.eu/en/news/european-medicines-agency-recommends-suspension-marketing-authorisation-acomplia.
8. Wilding J.P.H., Batterham R.L., Calanna S., Davies M., Van Gaal L.F., Lingvay I., et al. 'Once-Weekly Semaglutide in Adults with Overweight or Obesity'. *New England Journal of Medicine.* 2021;384(11):989–1002.
9. It is worth noting that, at the time of writing, Ozempic is not licensed for weight loss in the UK.
10. Holst J.J., Ørskov C., Vagn Nielsen O., Schwartz T.W. 'Truncated glucagon-like peptide I, an insulin-releasing hormone from the distal gut'. *FEBS Letters.* 1987;211(2):169–74.
11. Larsen P.J., Fledelius C., Bjerre Knudsen L., Tang-Christensen M. 'Systemic Administration of the Long-Acting GLP-1 Derivative NN2211 Induces Lasting and Reversible Weight Loss in Both Normal and Obese Rats'. *Diabetes.* 2001;50(11):2530–2539.
12. Mentlein R., Gallwitz B. and Schmidt W.E. 'Dipeptidyl-peptidase IV hydrolyses gastric inhibitory polypeptide, glucagon-like peptide -1(7–36)amide, peptide histidine methionine and is responsible for their degradation in human serum'. *European Journal of Biochemistry.* 1993;214:829–835.
13. Holst, J.J. *The Story of GLP-1.* FADL's. 2024.
14. Holst, J.J. *The Story of GLP-1.* FADL's. 2024. Chapter by Mads Krogsgaard Thomsen.
15. Pi-Sunyer X., Astrup A., Fujioka K., Greenway, F., et al. 'A Randomized, Controlled Trial of 3.0 mg of Liraglutide in Weight Management'. *New England Journal of Medicine.* 2015;373:11–22.
16. Jastreboff A.M., Aronne L.J., Ahmad N.N., Wharton S., Connery L., Alves B., et al. 'Tirzepatide Once Weekly for the Treatment of Obesity'. *New England Journal of Medicine.* 2022;387(3):205–16.
17. Tang-Christensen, M., Larsen, P.J., Göke, R., et al. 'Central administration of GLP-1-(7-36) amide inhibits food and water intake in rats'. *Am J Physiol.* 1996 Oct;271(4 Pt 2):R848–56.

18. Hristakeva S., Liaukonyte J., Feler L. 'The No-Hunger Games: How GLP-1 Medication Adoption is Changing Consumer Food Purchases.' *Cornell SC Johnson College of Business Research Paper.* December 2024.
19. Burn-Murdoch J. 'We may have passed peak obesity'. *Financial Times* [online]. 4.20.24.

# Part 2: SEX AND DRUG SUPERSTIMULI

## 9. Sex and the bottle's spell

1. Gwynne D.T., Rentz D.C.F. 'Beetles on the Bottle: Male Buprestids Mistake Stubbies for Females (Coleoptera)'. *Australian Journal of Entomology*. 1983;22(1):79–80.
2. Stirling J. 'University of Toronto Mississauga professor wins Ig Nobel Prize for beer, sex research'. 2011. https://www.eurekalert.org/news-releases/919856.
3. Danmarks Statistik. 'Antal enlige voksne har rundet 1,6 mio. Personer'. 2016. https://www.dst.dk/da/Statistik/nyheder-analyser-publ/nyt/NytHtml?cid=20988.
4. Rosenfeld M.J., Thomas R.J., Hausen S. 'Disintermediating your friends: How online dating in the United States displaces other ways of meeting'. *Proc Natl Acad Sci USA*. 2019;116(36):17753–8.
5. https://web.stanford.edu/~mrosenfe/
6. Iyengar S.S., Lepper M.R. 'When choice is demotivating: can one desire too much of a good thing?' *J Pers Soc Psychol.* 2000;79(6):995–1006.
7. D'Angelo J., Toma C. 'There Are Plenty of Fish in the Sea: The Effects of Choice Overload and Reversibility on Online Daters' Satisfaction With Selected Partners'. *Media Psychology*. 2016;20:1–27.
8. Uggla C., Andersson G. 'Higher divorce risk when mates are plentiful? Evidence from Denmark'. *Biology Letters*. 2018;14(9).
9. Weber G.W., Lukeneder A., Harzhauser M., Mitteroecker P., Wurm L., Hollaus L.-M., et al. 'The microstructure and the origin of the Venus from Willendorf'. *Scientific Reports*. 2022;12(1):2926.

10. Deaner R.O., Khera A.V., Platt M.L. 'Monkeys pay per view: adaptive valuation of social images by rhesus macaques'. *Curr Biol.* 2005;15(6):543–8.
11. Singh D. 'Adaptive significance of female physical attractiveness: Role of waist-to-hip ratio'. *Journal of Personality and Social Psychology.* 1993 Aug;65(2):293–307.
12. Beach F.A., Jordan L. 'Sexual Exhaustion and Recovery in the Male Rat'. *Quarterly Journal of Experimental Psychology.* 1956;8(3):121–33.
13. Wilson J.R., Kuehn R.E., Beach F.A. 'Modification in the sexual behavior of male rats produced by changing the stimulus female'. *J Comp Physiol Psychol.* 1963;56:63644.
14. Beamer W., Bermant G., Clegg M.T. 'Copulatory behaviour of the ram, Ovis aries. II: Factors affecting copulatory satiation'. *Animal Behaviour.* 1969;17(4):706–11.
15. Hughes S.M., Aung T., Harrison M.A., LaFayette J.N., Gallup G.G., Jr. 'Experimental Evidence for Sex Differences in Sexual Variety Preferences: Support for the Coolidge Effect in Humans'. *Arch Sex Behav.* 2021;50(2):495–509.
16. Schacht, R. and Kramer, K.L. 'Are We Monogamous? A Review of the Evolution of Pair-Bonding in Humans and Its Contemporary Variation Cross-Culturally'. *Front. Ecol. Evol.* 2019;7.
17. 'Chris Rock's porn addiction and cheating which wrecked his own marriage'. Wales Online. 29 March 2022. https://www.walesonline.co.uk/lifestyle/showbiz/chris-rocks-controversies-during-decades-23528752.
18. 'Terry Crews reveals impact his pornography addiction had on his relationship with wife and children'. *Independent* [online]. 25 July 2023. https://www.independent.co.uk/life-style/terry-crews-pornography-addiction-family-b2381723.html.
19. 'Billie Eilish says watching porn as a child "destroyed my brain"'. *Guardian* [online]. 15 December 2021. https://www.theguardian.com/music/2021/dec/15/billie-eilish-says-watching-porn-gave-her-nightmares-and-destroyed-my-brain.
20. International Classification of Diseases ERI-. 6C72 Compulsive sexual behaviour disorder. Impulse Control Disorders. World Health Organization. 2019/2021.

21. Kraus S.W., Voon V., Potenza M.N. 'Should compulsive sexual behavior be considered an addiction?' *Addiction*. 2016;111(12): 2097–106.
22. Als Ringheim H., Killerich K., et al. 'Pornografisk indhold påvirker danske børn og unge'. Mediesunhed for børn og unge. 2021.
23. Wellings K., Palmer M.J., Machiyama K., Slaymaker E. 'Changes in, and factors associated with, frequency of sex in Britain: evidence from three National Surveys of Sexual Attitudes and Lifestyles (Natsal)'. *BMJ*. 2019;365:l1525.
24. Twenge J.M., Sherman R.A., Wells B.E. 'Declines in Sexual Frequency among American Adults, 1989–2014'. *Arch Sex Behav*. 2017;46(8):2389–401.
25. Beutel M., Burghardt J., Tibubos A.N., Klein E., Schmutzer G., Brähler E. 'Declining Sexual Activity and Desire in Men – Findings From Representative German Surveys, 2005 and 2016'. *The Journal of Sexual Medicine*. 2018;15:750–6.
26. Twenge J.M., Sherman R.A., Wells B.E. 'Sexual Inactivity During Young Adulthood Is More Common Among U.S. Millennials and iGen: Age, Period, and Cohort Effects on Having No Sexual Partners After Age 18'. *Arch Sex Behav*. 2017;46(2):433–40.
27. Twenge J.M., Sherman R.A., Wells B.E. 'Declines in Sexual Frequency among American Adults, 1989–2014'. *Arch Sex Behav*. 2017;46(8):2389–401.
28. Miller D.J., McBain K.A., Li W.W., Raggatt P.T.F. 'Pornography, preference for porn-like sex, masturbation, and men's sexual and relationship satisfaction'. *Personal Relationships*. 2019;26(1):93–113.
29. Ibid.
30. Snowdon C.T., Tannenbaum P.L., Schultz-Darken N.J., Ziegler T.E., Ferris C.F. 'Conditioned sexual arousal in a nonhuman primate'. *Hormones and Behavior*. 2011;59(5):696–701.
31. Pfaus J.G., Erickson K.A., Talianakis S. 'Somatosensory conditioning of sexual arousal and copulatory behavior in the male rat: A model of fetish development'. *Physiology & Behavior*. 2013;122:1–7.

## 10. Shortcut to euphoria

1. Grotenhermen F. 'Pharmacokinetics and Pharmacodynamics of Cannabinoids'. *Clinical Pharmacokinetics*. 2003;42(4):327–60.
2. Ibid.
3. Rømer Thomsen K., Lindholst C., Thylstrup B., Kvamme S., Reitzel L.A., Worm-Leonhard M., et al. 'Changes in the composition of cannabis from 2000-2017 in Denmark: Analysis of confiscated samples of cannabis resin'. *Exp Clin Psychopharmacol.* 2019;27(4):402–11.
4. ElSohly M.A., Mehmedic Z., Foster S., Gon C., Chandra S., Church J.C. 'Changes in Cannabis Potency Over the Last 2 Decades (1995–2014): Analysis of Current Data in the United States'. *Biol Psychiatry*. 2016;79(7):613–9.
5. Chandra S., Radwan M.M., Majumdar C.G., Church J.C., Freeman T.P., ElSohly M.A. 'New trends in cannabis potency in USA and Europe during the last decade (2008–2017)'. *Eur Arch Psychiatry Clin Neurosci.* 2019;269(1):5–15.
6. 'Legal weed didn't deliver on its promises'. *Atlantic* [online]. 31 January 2025. https://www.theatlantic.com/ideas/archive/2025/01/marijuana-legalization-drawbacks/681519/.
7. Payne K.S., Mazur D.J., Hotaling J.M., Pastuszak A.W. 'Cannabis and Male Fertility: A Systematic Review'. *J Urol.* 2019;202(4):674–81.
8. Hedges J.C., Hanna C.B., Bash J.C., Boniface E.R., Burch F.C., Mahalingaiah S., et al. 'Chronic exposure to delta-9-tetrahydrocannabinol impacts testicular volume and male reproductive health in rhesus macaques'. *Fertil Steril.* 2022;117(4):698–707.
9. Hines L.A., Heron J., Zammit S. 'Incident psychotic experiences following self-reported use of high-potency cannabis: Results from a longitudinal cohort study'. *Addiction.* 2024;119(9):1629–34.
10. Marconi A., Di Forti M., Lewis C.M., Murray R.M., Vassos E. 'Meta-analysis of the Association Between the Level of Cannabis Use and Risk of Psychosis'. *Schizophrenia Bulletin.* 2016;42(5):1262–9.
11. Di Forti M., Quattrone D., Freeman T.P., Tripoli G., Gayer

-Anderson C., Quigley H., et al. 'The contribution of cannabis use to variation in the incidence of psychotic disorder across Europe (EU-GEI): a multicentre case-control study'. *Lancet Psychiatry*. 2019;6(5):427–36.

12. Freeman T.P., van der Pol P., Kuijpers W., Wisselink J., Das R.K., Rigter S., et al. 'Changes in cannabis potency and first-time admissions to drug treatment: a 16-year study in the Netherlands'. *Psychol Med.* 2018;48(14):2346–52.
13. Morgan C.J., Gardener C., Schafer G., Swan S., Demarchi C., Freeman T.P., et al. 'Sub-chronic impact of cannabinoids in street cannabis on cognition, psychotic-like symptoms and psychological well-being'. *Psychol Med.* 2012;42(2):391–400.
14. Hjorthøj C., Posselt C.M., Nordentoft M. 'Development Over Time of the Population-Attributable Risk Fraction for Cannabis Use Disorder in Schizophrenia in Denmark'. *JAMA Psychiatry*. 2021;78(9):1013–9.
15. Keerthy D., Chandan J.S., Abramovaite J., Gokhale K.M., Bandyopadhyay S., Day E., et al. 'Associations between primary care recorded cannabis use and mental ill health in the UK: a population-based retrospective cohort study using UK primary care data'. *Psychological Medicine*. 2023;53(5):2106–15.
16. Moore M.H., Gerstein D.R., editors. *Alcohol and Public Policy: Beyond the Shadow of Prohibition*. National Research Council Panel on Alternative Policies Affecting the Prevention of Alcohol Abuse and Alcoholism. National Academies Press (US). 1981.
17. Law M.T., Marks M.S. 'Did Early Twentieth-Century Alcohol Prohibition Affect Mortality?' *Economic Inquiry*. 2020;58(2):680–97.
18. Blocker J.S. 'Did Prohibition Really Work? Alcohol Prohibition as a Public Health Innovation'. *American Journal of Public Health*. 2006;96(2):233–43.
19. Taylor W., McCarthy D., Wilbur K.C. 'Online Gambling Policy Effects on Tax Revenue and Irresponsible Gambling'. 2024; SMU Cox School of Business Research Paper No. 24-7. https://ssrn.com/abstract=4856684 or http://dx.doi.org/10.2139/ssrn.4856684.
20. Hollenbeck B., Larsen P., Proserpio D. 'The Financial Consequences of Legalized Sports Gambling'. SSRN. 2024. https://papers.ssrn.com/sol3/papers.cfm?abstract_id=4903302.

21. Rubin-Kahana D.S., Crépault J.-F., Matheson J., Le Foll B. 'The impact of cannabis legalization for recreational purposes on youth: A narrative review of the Canadian experience'. *Frontiers in Psychiatry*. 2022;13.
22. NIDA. 'Marijuana and hallucinogen use among young adults reached all time-high in 2021'. nida.nih.gov. 2022. https://nida.nih.gov/news-events/news-releases/2022/08/marijuana-and-hallucinogen-use-among-young-adults-reached-all-time-high-in-2021.
23. Zellers S.M., Ross J.M., Saunders G.R.B., Ellingson J.M., Anderson J.E., Corley R.P., et al. 'Impacts of recreational cannabis legalization on cannabis use: a longitudinal discordant twin study'. *Addiction*. 2023;118(1):110–8.
24. Ibid.
25. Myran D.T., Cantor N., Finkelstein Y., Pugliese M., Guttmann A., Jesseman R., et al. 'Unintentional Pediatric Cannabis Exposures After Legalization of Recreational Cannabis in Canada'. *JAMA Netw Open*. 2022;5(1):e2142521.
26. Amissah R.Q., Vogt N.A., Chen C., Urban K., Khokhar J. 'Prevalence and characteristics of cannabis-induced toxicoses in pets: Results from a survey of veterinarians in North America'. *PLoS One*. 2022;17(4):e0261909.
27. Jayawardhana J., Hou J., Freeman P., Talbert J.C. 'Association of State Cannabis Legalization With Cannabis Use Disorder and Cannabis Poisoning'. *JAMA Psychiatry*. 23.12.24.
28. Farmer C.M., Monfort S.S., Woods A.N. 'Changes in Traffic Crash Rates After Legalization of Marijuana: Results by Crash Severity'. *Journal of Studies on Alcohol and Drugs*. 2022;83(4):494–501.
29. Marinello S., Powell L.M. 'The impact of recreational cannabis markets on motor vehicle accident, suicide, and opioid overdose fatalities'. *Social Science & Medicine*. 2023;320:115680.
30. Holdway J. 'Car crash deaths involving cannabis increasing and more likely to involve alcohol'. Boston Medical Center. 2021.
31. Tolan N.V., Terebo T., Chai P.R., Erickson T.B., Hayes B.D., Uljon S.N., et al. 'Impact of marijuana legalization on cannabis-related visits to the emergency department'. *Clin Toxicol (Phila)*. 2022;60(5):585–95.

32. Roberts B.A. 'Legalized Cannabis in Colorado Emergency Departments: A Cautionary Review of Negative Health and Safety Effects'. West J Emerg Med. 2019;20(4):557–72.
33. 'Cannabis'. Tobacco Tactics, University of Bath. 31 March 2023. (https://www.tobaccotactics.org/article/cannabis/).

## 11. The happiness plant

1. Booth M. *Opium: A History*. Thomas Dunne Books. 1998. p.380.
2. Salavert A., Zazzo A., Martin L., Antolín F., Gauthier C., Thil F., et al. 'Direct dating reveals the early history of opium poppy in western Europe'. *Scientific Reports*. 2020;10(1):20263.
3. Weid M., Ziegler J. and Kutchan T.M. 'The roles of latex and the vascular bundle in morphine biosynthesis in the opium poppy, *Papaver somniferum*'. *PNAS*. 2004;101(38):13957–13962.
4. Offit, P. 'God's Own Medicine'. *Skeptical Inquirer.* 2017;41(2). https://skepticalinquirer.org/2017/03/gods-own-medicine/.
5. Martin Booth. *Opium: A History.* St Martin's Griffin. 1999. p.172.
6. Offit, P. 'God's Own Medicine'. *Skeptical Inquirer.* 2017;41(2). https://skepticalinquirer.org/2017/03/gods-own-medicine/.
7. Radden Keefe, Patrick. *Empire of Pain: The Secret History of the Sackler Dynasty.* Doubleday, 2021. p.221.
8. 'Understanding the Opioid Overdose Epidemic'. CDC Overdose Prevention. 1 November 2024. https://www.cdc.gov/overdose-prevention/about/understanding-the-opioid-overdose-epidemic.html.
9. https://www.govinfo.gov/content/pkg/CHRG-116hhrg43010/html/CHRG-116hhrg43010.htm.
10. https://www.docketbird.com/court-documents/In-re-National-Prescription-Opiate-Litigation/Wright-Curtis-Purdue-2018-12-19-Transcript/ohnd-1:2017-md-02804-02177-020, pp.25–26.
11. Radden Keefe, Patrick. *Empire of Pain: The Secret History of the Sackler Dynasty.* Doubleday, 2021. p.211.
12. Radden Keefe, Patrick. *Empire of Pain: The Secret History of the Sackler Dynasty.* Doubleday, 2021. p.214.
13. *Commonwealth of Massachusetts* v. *Purdue Pharma L.P.*, First Amended Complaint, paragraph 241.

14. Radden Keefe, Patrick. *Empire of Pain: The Secret History of the Sackler Dynasty.* Doubleday, 2021. pp.293–294.
15. O'Connor A. 'Coca-Cola Funds Scientists Who Shift Blame for Obesity Away From Bad Diets'. *The New York Times* [online]. 9.8.15. https://archive.nytimes.com/well.blogs.nytimes.com/2015/08/09/coca-cola-funds-scientists-who-shift-blame-for-obesity-away-from-bad-diets/.
16. Cook P. *Paying the Tab: The Costs and Benefits of Alcohol Control.* Princeton University Press. 2011.
17. Maremont M., Berzon A. 'How Often Do Gamblers Really Win?' *Wall Street Journal* [online]. 11.10.13. https://www.wsj.com/articles/how-often-do-gamblers-really-win-1381514164.

## 12. Dopamine and what you need to know about it

1. Olds J,. Milner P. 'Positive reinforcement produced by electrical stimulation of septal area and other regions of rat brain'. *Journal of Comparative and Physiological Psychology.* 1954;47(6):419–27.
2. Routtenberg A., Lindy J. 'Effects of the availability of rewarding septal and hypothalamic stimulation on bar pressing for food under conditions of deprivation'. *Journal of Comparative and Physiological Psychology.* 1965;60(2):158–61.
3. Black J., Belluzzi J.D., Stein L. 'Reinforcement delay of one second severely impairs acquisition of brain self-stimulation'. *Brain Research.* 1985;359(1):113–9.
4. Heath R.G. 'Pleasure and pain activity in man. Deep and surface electroencephalograms during orgasm'. *J Nerv Ment Dis.* 1972 Jan;154(1):3–18.
5. Portenoy R.K., Jarden J.O., Sidtis J.J., Lipton R.B., Foley K.M., Rottenberg D.A. 'Compulsive thalamic self-stimulation: a case with metabolic, electrophysiologic and behavioral correlates'. *Pain.* 1986;27(3):277–90.
6. Berridge K.C., Kringelbach M.L. 'Pleasure systems in the brain'. *Neuron.* 2015;86(3):646–64.
7. Szczypka M.S., Rainey M.A., Kim D.S. et al. 'Feeding behavior in dopamine-deficient mice'. *Proc Natl Acad Sci USA.* 1999 Oct 12;96(21):12138–12143.

8. Cannon C.M. and Palmiter R.D. 'Reward without Dopamine'. *Journal of Neuroscience.* 26 November 2003, 23 (34) 10827–10831.
9. Politis M., Loane C., Wu K., O'Sullivan S.S., Woodhead Z., Kiferle L., et al. 'Neural response to visual sexual cues in dopamine treatment-linked hypersexuality in Parkinson's disease'. *Brain.* 2013;136(2):400–11.
10. Weintraub D., Koester J., Potenza M.N., Siderowf A.D., Stacy M., Voon V., et al. 'Impulse Control Disorders in Parkinson Disease: A Cross-Sectional Study of 3090 Patients'. *Archives of Neurology.* 2010;67(5):589–95.
11. Fermaglich J. 'The Case of the Frozen Addicts'. *JAMA.* 1996;275(5):407–8.
12. Schultz W., Dayan P. and Montague P.R. 'A Neural Substrate of Prediction and Reward'. *Science.* 1997, 275,1593–1599.
13. Robins L.N., Davis D.H., Nurco D.N. 'How permanent was Vietnam drug addiction?' *Am J Public Health.* 1974;64 Suppl 12(12 Suppl):38–43.
14. Hendershot C.S., Wardell J.D., Samokhvalov A.V., Rehm J. 'Effects of naltrexone on alcohol self-administration and craving: meta-analysis of human laboratory studies'. *Addict Biol.* 2016 Jul 14;22(6):1515–1527.
15. Harris R.C. and Zhang M. 'Dopamine, the Kidney, and Hypertension'. *Curr Hypertens Rep.* 2012 Apr;14(2):138–143.
16. Matt S.M. and Gaskill P.J. 'Where Is Dopamine and how do Immune Cells See it?: Dopamine-Mediated Immune Cell Function in Health and Disease'. *Journal of Neruoimmune Pharmacology.* 2020; 15:114–164.
17. Bucolo C., Marco Leggio G., Drago F., Salomone S. 'Dopamine outside the brain: The eye, cardiovascular system and endocrine pancreas'. *Pharmacology & Therapeutics.* Nov 2019 Vol 203.

## Part 3: SCREEN SUPERSTIMULI

### 13. *Friends*, just in real life

1. 'It's All A/Bout Testing: The Netflix Experimentation Platform'. Medium. 29 April 2016. https://netflixtechblog.com/its-all-a-bout-testing-the-netflix-experimentation-platform-4e1ca458c15.

2. Segal D. 'A Kid's Show Juggernaut That Leaves Nothing to Chance'. *The New York Times* [online]. 5.5.22. https://www.nytimes.com/2022/05/05/arts/television/cocomelon-moonbug-entertainment.html.
3. 2024 Global Internet Phenomena Report. https://www.sandvine.com/phenomena.
4. 'Netflix's biggest competitor? Sleep'. *Guardian* [online]. 18 April 2017. https://www.theguardian.com/technology/2017/apr/18/netflix-competitor-sleep-uber-facebook.
5. Horton D. and Wohl R.R. 'Mass Communication and Para-Social Interaction: Observations on Intimacy at a Distance.' *Psychiatry*. 1956;9(3):215–229.
6. Gerace A. 'When TV neighbours become good friends: Understanding *Neighbours* fans' feelings of grief and loss at the end of the series'. *PLoS One*. 2024 Jun 12;19(6):e0302160.
7. Haak W., Lazaridis I., Patterson N., Rohland N., Mallick S., Llamas B., et al. 'Massive migration from the steppe was a source for Indo-European languages in Europe'. *Nature*. 2015;522(7555):207–11.
8. Haak W., Lazaridis I., Patterson N. et al. 'Massive migration from the steppe was a source for Indo-European languages in Europe'. *Nature*. 2015;522:201–211.
9. Fortes-Lima C., Burgarella C., et al. 'The genetic legacy of the expansion of Bantu-speaking peoples in Africa'. *Nature*. 2023;625(7995):1–8.
10. Wilson M. and Daly M. 'Competitiveness, risk taking, and violence: the young male syndrome'. *Ethology and Sociobiology*. 1985;6(1): 57–73.
11. Lester D., Yang B., Lindsay M. 'Suicide Bombers: Are Psychological Profiles Possible?' *Studies in Conflict & Terrorism*. 2004;27(4):283–295.
12. Sela-Shayovitz R. 'Suicide Bombers in Israel: Their Motivations, Characteristics, and Prior Activity in Terrorist Organizations'. *IJCV*. 2007;1(2):160–8.
13. Williams A.F. and Shabanova V.I. 'Responsibility of drivers, by age and gender, for motor-vehicle crash deaths'. *J Safety Res*. 2003;34(5):527–31.
14. Jeanmonod R.K., Jeanmonod D., Damewood S., Perry C., Powers M., Lazansky V. 'Punch Injuries: Insights into Intentional Closed Fist Injuries'. *West J Emerg Med*. 2011 Feb;12(1):6–10.

15. Przybylski A., Weinstein N. 'Violent video game engagement is not associated with adolescents' aggressive behaviour: evidence from a registered report'. *R Soc Open Sci*. 2019;13;6(2):171474.
16. Kuhn S., Kugler D., et al. 'Does playing violent video games cause aggression? A longitudinal intervention study'. *Mol Psychiatry*. 2019 24:1220–34.
17. Danmarks Statistik. 'Færre 18-årige dømmes efter straffeloven'. dst.dk. 2023. https://www.dst.dk/da/Statistik/nyheder-analyser-publ/nyt/NytHtml?cid=46153.
18. Beerthuizen M., Weijters G., van der Laan, A. 'The release of Grand Theft Auto V and registered juvenile crime in the Netherlands'. *European Journal of Criminology*. 2017;14(6):751–65.
19. Dahl G., DellaVigna S. 'Does Movie Violence Increase Violent Crime?' *The Quarterly Journal of Economics*. 2009;124;2:677–734.
20. Ottosen M.H., Andreasen A.G., Dahl K.M., Lausten M, Rayce S.B., Tagmose B.B. *Børn og unge i Danmark: Velfærd og trivsel 2022*. VIVE – Det Nationale Forsknings- og Analysecenter for Velfærd. 2022. https://www.vive.dk/da/udgivelser/boern-og-unge-i-danmark-velfaerd-og-trivsel-2022-0xgg53xk/.
21. Sharkey P. 'Homebound: The Long-Term Rise in Time Spent at Home Among U.S. Adults'. *Sociological Science*. Aug 2024 10.15195/v11.a20.

## 14. Scrolling into infinity

1. Bryson A., MacKerron G. 'Are You Happy While You Work?' *The Economic Journal*. 2017;127(599):106–25.
2. Taylor P. 'Number of smartphone mobile network subscriptions worldwide from 2016 to 2022, with forecasts from 2023 to 2028'. statistica.com. 2022. https://www.statista.com/statistics/330695/number-of-smartphone-users-worldwide/.
3. United Nations Children's Fund and the World Health Organization. *State of the World's Sanitation: An urgent call to transform sanitation for better health, environments, economies and societies*. 2020.
4. Our World in Data. 'Daily hours spent with digital media in the United States'. BOND Internet Trends (2019). 2019.

https://ourworldindata.org/grapher/daily-hours-spent-with-digital-media-per-adult-user.

5. Kemp S. 'Digital 2022: Global Overview Report'. DataReportal. 2022. https://datareportal.com/reports/digital-2022-global-overview-report.
6. Pew Research Center. 'Teens, social media and technology 2022'. [online]. 2022. https://www.pewresearch.org/internet/2022/08/10/teens-social-media-and-technology-2022/pj_2022-08-10_teens-and-tech_0-06b/.
7. Backlinko. 'Revealing Average Screen Time Statistics'. Backlinko.com. 11.3.24. https://backlinko.com/screen-time-statistics?utm_source=chatgpt.com.
8. Ferster C.B. and Skinner B.F. 'Schedules of Reinforcement'. 1957. https://www.bfskinner.org/wp-content/uploads/2015/05/Schedules_of_Reinforcement_PDF.pdf.
9. Allen M. 'Sean Parker unloads on Facebook: "God only knows what it's doing to our children's brains"'. Axios. 9.11.17. https://www.axios.com/2017/12/15/sean-parker-unloads-on-facebook-god-only-knows-what-its-doing-to-our-childrens-brains-1513306792.
10. Samrai Y. 'How Stanford Profits Off Addiction'. *The Stanford Review* [online]. 4.2.20. https://stanfordreview.org/how-stanford-profits-tech-addiction-social-media/.
11. Srinivisan K. 'Paying attention'. University of Chicago Booth School of Business. 2023. https://karthikecon.github.io/karthiksrinivasan.org/paying_attention.pdf.
12. Knowles T. 'I'm so sorry, says inventor of endless online scrolling'. *The Times*. 27.4.19. https://www.thetimes.co.uk/article/i-m-so-sorry-says-inventor-of-endless-online-scrolling-9lrv59mdk.
13. 'Former Facebook executive: social media is ripping society apart'. *Guardian* [online]. 12 December 2017. https://www.theguardian.com/technology/2017/dec/11/facebook-former-executive-ripping-society-apart.
14. McNamee, R. *Zucked: Waking Up to the Facebook Catastrophe*. Penguin Press. 2019.
15. 'The person behind the Like button says software is wasting our time'. *The Verge*. March 2018. https://www.theverge.com/2018/3/28/17172404/justin-rosenstein-asana-social-media-facebook-timeline-gantt.

16. Bowles N. 'Silicon Valley Nannies Are Phone Police for Kids'. *The New York Times*. 26.10.18. https://www.nytimes.com/2018/10/26/style/silicon-valley-nannies.html.
17. Bowles N. 'A Dark Consensus About Screens and Kids Begins to Emerge in Silicon Valley'. *The New York Times*. 26.10.18. https://www.nytimes.com/2018/10/26/style/phones-children-silicon-valley.html – comments container.
18. Fennell J.G., Talas L., Baddeley R.J., Cuthill I.C., Scott-Samuel N.E. 'Optimizing colour for camouflage and visibility using deep learning: the effects of the environment and the observer's visual system'. *J R Soc Interface*. 2019;16(154):20190183.
19. Zahavi A.A. *The Handicap Principle: A Missing Piece of Darwin's Puzzle*. Oxford University Press. 1999.
20. Beknazar-Yuzbashev G., Jiménez Durán R., McCrosky J., Stalinski M. 'Toxic Content and User Engagement on Social Media: Evidence from a Field Experiment'. SSRN. 2023;82.

## 15. A little fish in a big pond

1. Team USA. Michael Phelps: Athlete Bio. teamusa.org. 2023. https://www.teamusa.com/profiles/michael-phelps.
2. Lätt E., Jürimäe J., Mäestu J., Purge P., Rämson R., Haljaste K., et al. 'Physiological, biomechanical and anthropometrical predictors of sprint swimming performance in adolescent swimmers'. *J Sports Sci Med*. 2010;9(3):398–404.
3. Jürimäe J., Haljaste K., Cicchella A., Lätt E., Purge P., Leppik A., et al. 'Analysis of swimming performance from physical, physiological, and biomechanical parameters in young swimmers'. *Pediatr Exerc Sci*. 2007;19(1):70–81.
4. 'Michael Phelps'. United States Olympic and Paralympic Museum. https://usopm.org/michael-phelps/.
5. 'Michael Phelps: Going Five Years Without Missing A Single Day Of Training'. Forbes. https://www.forbes.com/video/5180394324001/michael-phelps-going-five-years-without-missing-a-single-day-of-training/.
6. Gladwell M. *David and Goliath: Underdogs, Misfits, and the Art of Battling Giants*. Little, Brown and Company. 2013;81–94.

7. Marsh H.W., Hau K.T. 'Big-fish-little-pond effect on academic self -concept. A cross-cultural (26-country) test of the negative effects of academically selective schools'. *Am Psychol.* 2003;58(5):364–76.
8. Seaton M., Marsh H.W., Craven R.G. 'Big-Fish-Little-Pond Effect: Generalizability and Moderation – Two Sides of the Same Coin'. *American Educational Research Journal.* 2010;47(2):390–433.
9. Seaton M., Marsh H., Craven R. 'Earning Its Place as a Pan-Human Theory: Universality of the Big-Fish-Little-Pond Effect Across 41 Culturally and Economically Diverse Countries'. *Journal of Educational Psychology.* 2009;101:403–19.
10. Chanal J.P., Marsh H.W., Sarrazin P.G., Bois J.E. 'Big-Fish-Little-Pond Effects on Gymnastics Self-Concept: Social Comparison Processes in a Physical Setting'. *Journal of Sport and Exercise Psychology.* 2005;27(1):53–70.
11. Marsh H.W., Morin A.J.S., Parker P.D. 'Physical Self-Concept Changes in a Selective Sport High School: A Longitudinal Cohort-Sequence Analysis of the Big-Fish-Little-Pond Effect'. *Journal of Sport and Exercise Psychology.* 2015;37(2):150–63.
12. Simler K. 'Social Status: Down the Rabbit Hole'. 2015. https://meltingasphalt.com/social-status-down-the-rabbit-hole/.
13. Wrangham R. *The Goodness Paradox: The Strange Relationship Between Virtue and Violence in Human Evolution.* Pantheon. 2019;165.
14. Sundhedsstyrelsen. *Prævalens, incidens og aktivitet i sundhedsvæsenet for børn og unge med angst eller depression, ADHD og spiseforstyrrelse.* Sundhedstyrelsen. 2017.
15. Rosendahl Jensen H.A., Davidsen M., Rossen Møller S., Ellegaard Ibáñez Román J., Kragelund K., Illemann Christensen A., Ekholm O. *Den Nationale Sundhedsprofil 2021.* Sundhedsstyrelsen. 2022.
16. Dooley B., O'Connor C., Fitzgerald A., O'Reilly A. *The National Study of Youth and Mental Health in Ireland.* MyWorldSurvey.ie: University College Dublin School of Psychology. 2012.
17. Beckman L., Hellström L. 'Views on Adolescents' Mental Health in Sweden-A Qualitative Study among Different Professionals Working with Adolescents'. *Int J Environ Res Public Health.* 2021;18(20).
18. DREES. 'Crise sanitaire: hausse des syndromes dépressifs et des consultations pour ce motif'. 12.3.21. https://drees.solidarites-sante.gouv.fr/communique-de-presse/crise-sanitaire-hausse-des-syndromes-depressifs-et-des-consultations-pour-ce.

19. Twenge J. 'How Much Is Social Media to Blame for Teens' Declining Mental Health?' Institute for Family Studies [Internet]. Blog post. 11.4.22. https://ifstudies.org/blog/how-much-is-social-media-to-blame-for-teens-declining-mental-health.
20. Shaw N. 'Self-harm hospital admissions for teenage girls triple in a decade'. Walesonline.co.uk. 1.2.21. https://www.walesonline.co.uk/news/uk-news/self-harm-hospital-admissions-teenage-19740803.
21. Arenas-Arroyo E., Fernández-Kranz D, Nollenberger N. 'High Speed Internet and the Widening Gender Gap in Adolescent Mental Health: Evidence from Hospital Records'. IZA Institute for Labor Economics. 2022.
22. Braghieri L., Levy R., Makarin A. 'Social Media and Mental Health'. *American Economic Review*. 2022;112(11):3660–93.
23. Allcott H., Braghieri L., Eichmeyer S., Gentzkow M. 'The Welfare Effects of Social Media'. *American Economic Review*. 2020;110(3): 629–76.
24. Pedersen J., Rasmussen M.G.B., Sørensen S.O., Mortensen S.R., Olesen L.G., Brage S., et al. 'Effects of limiting digital screen use on well-being, mood, and biomarkers of stress in adults'. *Npj Mental Health Research*. 2022;1(1):14.
25. Stieger S., Lewetz D. 'A Week Without Using Social Media: Results from an Ecological Momentary Intervention Study Using Smartphones'. *Cyberpsychology, Behavior, and Social Networking*. 2018;21(10):618–24.
26. 'The Fundamental Flaws of The Only Meta-Analysis of Social Media Reduction Experiments (And Why It Matters), Part 1'. After Babel. 29 August 2024. https://www.afterbabel.com/p/the-case-for-causality-part-1.
27. Deri S., Davidai S., Gilovich T. 'Home alone: why people believe others' social lives are richer than their own'. *Journal of Personality and Social Psychology*. 2017 Dec;113(6):858–877.
28. Eriksen I.M., Bakken A., Soest von T. 'Stress, press og psykiske plager blant unge 2020'. https://www.ungdata.no/stress-press-og-psykiske-plager-blant-unge/.
29. Robertson C.E., Pröllochs N., Schwarzenegger K., et al. 'Negativity drives online news consumption'. *Nat Hum Behav*. 2023;7:812–22.
30. Griffiths S., Harris E.A., Whitehead G., et al. 'Does TikTok contribute to eating disorders? A comparison of the TikTok

algorithms belonging to individuals with eating disorders versus healthy controls'. *Body Image.* 2024(51):101807.

31. Marsh, Sarah. 'TikTok investigating videos promoting starvation and anorexia'. 7 Decemeber 2020. *Guardian* [online]. https://www.theguardian.com/technology/2020/dec/07/tiktok-investigating-videos-promoting-starvation-and-anorexia.

## 16. From cosmetics to nuclear weapons

1. Safran R.J., et al. 'Dynamic Paternity Allocation as a Function of Male Plumage Color in Barn Swallows.' *Science.* 2005;309, 2210–12.
2. Safran R.J., et al. 'Sexual signal exaggeration affects physiological state in male barn swallows'. *Current Biology.* 2008;18;11:461–2.
3. Langlois J.H., Kalakanis L., Rubenstein A.J., Larson A., Hallam M., Smoot M. 'Maxims or myths of beauty? A meta-analytic and theoretical review'. *Psychol Bull.* 2000;126(3):390–423.
4. Coetzee V., Greeff J.M., Stephen I.D., Perrett D.I. 'Cross-cultural agreement in facial attractiveness preferences: the role of ethnicity and gender'. *PLoS One.* 2014;9(7):e99629.
5. Russell R. 'A Sex Difference in Facial Contrast and its Exaggeration by Cosmetics'. *Perception.* 2009;38(8), 1211–19.
6. Hammermesh D., Biddle J. 'Beauty and the Labor Market'. *American Economic Review.* 1993;84:1174–94.
7. Sundhedsdatastyrelsen. Plastikoperationer (esundhed.dk). https://www.esundhed.dk/Emner/Operationer-og-diagnoser/Plastikoperationer.
8. 'Zimmer-Gembeck M.J., Webb H.J., Farrell L.J. and Waters A.M. 'Girls' and boys' trajectories of appearance anxiety from age 10 to 15 years are associated with earlier maturation and appearance-related teasing'. *Dev Psychopathol.* 2018;30(1):337–350.

## 17. The ideal body on steroids

1. Frederick D.A. and Haselton M.G. 'Why Is Muscularity Sexy? Tests of the Fitness Indicator Hypothesis'. *Personality and Social Psychology Bulletin.* 2007;33(8)1167–1183.

2. Samuel E. 'Arnold Schwarzenegger Explains His Olympia-Era PED Use'. *Men'sHealth* [online]. 24.5.23. https://www.menshealth.com/fitness/a43944437/arnold-schwarzenegger-steroid-use-bodybuilding/.
3. The Barbell Team. 'Dolph Lundgren Admits to Taking Steroids'. TheBarbell.com. 11.5.23. https://www.thebarbell.com/dolph-lundgren-steroids/.
4. Reuters. 'Stallone pleads guilty in hormone import case'. Reuters.com. 9.8.07. https://www.reuters.com/article/lifestyle/stalone-pleads-guilty-in-hormone-import-case-idUSSYD39413/.
5. 'Rappers Implicated In Steroids Scandal'. CityNews Everywhere. 14 January 2008. https://toronto.citynews.ca/2008/01/14/rappers-implicated-in-steroids-scandal/?.
6. Bhasin S., Storer T.W., Berman N., Callegari C., Clevenger B., Phillips J., et al. 'The Effects of Supraphysiologic Doses of Testosterone on Muscle Size and Strength in Normal Men'. *New England Journal of Medicine*. 1996;335(1):1–7.
7. Pope Jr H.G., Olivardia R., Gruber A., Borowiecki J. 'Evolving ideals of male body image as seen through action toys'. *International Journal of Eating Disorders*. 1999;26(1):65–72.
8. Xue L. and Perrett D. 'Misperceptions of opposite-sex preferences for thinness and muscularity'. *Br. J. Psychol.* 2021;112:247–264.
9. Frederick D.A., Fessler D.M.T., Haselton M.G. 'Do representations of male muscularity differ in men's and women's magazines?' *Body Image*. 2005;2(1):81–6.
10. Brada T. 'Leg-lengthening: The people having surgery to be a bit taller'. BBC.com. 5.12.20. https://www.bbc.com/news/world-55146906.
11. Soo Z. 'China keeping 1 hour daily limit on kids' online games'. APNews.com. 20.1.23. https://apnews.com/article/gaming-business-children-00db669defcc8e0ca1fc2dc54120a0b8.
12. Ye J. 'China looks to limit children to two hours a day on their phones'. Reuters.com. 3.8.23. https://www.reuters.com/world/china/china-issues-draft-guidelines-tighten-limits-use-apps-by-minors-2023-08-02/.

## Epilogue: A little human in a big world

1. Kanigel R. *The Man Who Knew Infinity: A Life of the Genius Ramanujan*. Washington Square Press. 2013.
2. 'The Grandmaster Experiment'. *Psychology Today* [online]. 1 July 2005. https://www.psychologytoday.com/au/articles/200507/the-grandmaster-experiment.
3. Raichle M.E. and Gusnard D.A. 'Appraising the brain's energy budget'. *Proc Natl Acad Sci USA*. 2002 Jul 29;99(16):10237–10239.
4. Kurzban R., Duckworth A., Kable J.W., Myers J. 'Cost-benefit models as the next, best option for understanding subjective effort'. *Behavioral and Brain Sciences*. 2013;36(6):707–26.
5. Wu R., Ferguson A.M., Inzlicht M. 'Do humans prefer cognitive effort over doing nothing?' *Journal of Experimental Psychology: General*. 2023;152(4):1069–79.
6. Klausen M.K., Thomsen M., Wortwein G., Fink-Jensen A. 'The role of glucagon-like peptide 1 (GLP-1) in addictive disorders'. *British Journal of Pharmacology*. 2022;179(4):625–41.
7. Martinelli S., Mazzotta Al., Longaroni M., Petrucciani N. 'Potential role of glucagon-like peptide-1 (GLP-1) receptor agonists in substance use disorder: A systematic review of randomized trials'. *Drug and Alcohol Dependence*. 2024 Nov;264:112424.

# Index